# LATCHUP IN CMOS TECHNOLOGY

# THE KLUWER INTERNATIONAL SERIES IN ENGINEERING AND COMPUTER SCIENCE

## VLSI, COMPUTER ARCHITECTURE AND DIGITAL SIGNAL PROCESSING

*Consulting Editor*

Jonathan Allen

**Other books in the series:**

*Logic Minimization Algorithms for VLSI Synthesis.* R.K. Brayton, G.D. Hachtel, C.T. McMullen, and A.L. Sangiovanni-Vincentelli. ISBN 0-89838-164-9.

*Computer-Aided Design and VLSI Device Development.* K.M. Cham, S.-Y. Oh, D. Chin, and J.L. Moll. ISBN 0-89838-204-1.

*Adaptive Filters: Structures, Algorithms, and Applications.* M.L. Honig and D.G. Messerschmitt. ISBN 0-89838-163-0.

*Introduction to VLSI Silicon Devices: Physics, Technology and Characterization.* B. El-Kareh and R.J. Bombard. ISBN 0-89838-210-6.

# LATCHUP IN CMOS TECHNOLOGY

## The Problem and Its Cure

by

**Ronald R. Troutman**
IBM Corporation

KLUWER ACADEMIC PUBLISHERS
Boston / Dordrecht / Lancaster

Distributors for North America:
Kluwer Academic Publishers
190 Old Derby Street
Hingham, Massachusetts 02043, USA

Distributors for the UK and Ireland:
Kluwer Academic Publishers
MTP Press Limited
Falcon House, Queen Square
Lancaster LA1 1RN, UNITED KINGDOM

Distributors for all other countries:
Kluwer Academic Publishers Group
Distribution Centre
Post Office Box 322
3300 AH Dordrecht, THE NETHERLANDS

---

Library of Congress Cataloging-in-Publication Data

Troutman, Ronald R.
  Latchup in CMOS technology.

  (The Kluwer international series in engineering
and computer science. VLSI, computer architecture,
and digital signal processing)
  Bibliography: p.
  Includes index.
  1. Metal oxide semiconductors, Complementary—
Defects.  2. Metal oxide semiconductors, Complementary—
Reliability.  I. Title.  II. Series.
TK7871.99.M44T76  1986      621.3815'2       86-2925
ISBN 0-89838-215-7

---

Copyright © 1986 by Kluwer Academic Publishers

All rights reserved. No part of this publication may be reproduced, stored in a retrieval system, or transmitted in any form or by any means, mechanical, photocopying, recording, or otherwise, without written permission of the publisher, Kluwer Academic Publishers, 190 Old Derby Street, Hingham, Massachusetts 02043.

Printed in the United States of America

# Dedication

To Gail, who picked up the slack in so many areas and provided substantial encouragement for, and able assistance in, writing and editing this book.

# CONTENTS

| | |
|---|---|
| List of Figures | xi |
| Preface: Why a Book on Latchup? | xv |
| Acknowledgements | xix |

| | |
|---|---|
| **Chapter 1. INTRODUCTION** | 1 |
| 1.1 CMOS: The Emerging VLSI Technology | 1 |
| 1.2 Latchup Prevention : A Historical Perspective | 3 |
| 1.3 Latchup in VLSI CMOS: A Continuing Concern | 5 |
| **Chapter 2. LATCHUP OVERVIEW** | 7 |
| 2.1 Structural Origins and a Lumped Element Model | 7 |
| 2.2 An Example: Latchup in a Floating N-well Structure | 15 |
| **Chapter 3. PROBLEM DESCRIPTION** | 23 |
| 3.1 Blocking and Latched States | 23 |
| 3.2 Conditions Necessary for Latchup | 24 |
| 3.3 Triggering Modes | 24 |
|     3.3.1 Output Node Overshoot/Undershoot | 25 |
|     3.3.2 Input Node Overshoot/Undershoot | 26 |
|     3.3.3 Avalanching N-well Junction | 26 |
|     3.3.4 Punchthrough from N-well to External N-Type Diffusion | 28 |
|     3.3.5 Punchthrough from Substrate to Internal P-Type Diffusion | 29 |
|     3.3.6 Parasitic Field Devices | 30 |
|     3.3.7 Photocurrent | 32 |
|     3.3.8 Avalanching Source/Drain Junctions | 33 |
|     3.3.9 Displacement Current | 33 |
| 3.4 Triggering Taxonomy | 34 |
|     3.4.1 Type 1: External Initiation of First Bipolar | 34 |
|     3.4.2 Type 2: Normal Bypass Current Initiation of Both Bipolars | 34 |
|     3.4.3 Type 3: Degraded Bypass Current Initiation of Both Bipolars | 35 |
|     3.4.4 Latchup Sequence and Classification Summary | 35 |
| **Chapter 4. LATCHUP MODELS AND ANALYSES** | 37 |
| 4.1 Antecedents of Latchup Models | 38 |
|     4.1.1 Hook Collector Transistor | 38 |
|     4.1.2 Semiconductor Controlled Rectifier (SCR) | 40 |
| 4.2 Evolution of the PNPN Lumped Element Model | 47 |
|     4.2.1 Resistive Elements | 48 |
|     4.2.2 Parasitic Bipolar Behavior | 49 |
|     4.2.3 Previous Latchup Criteria | 51 |
| 4.3 Physical Analysis of Latchup: A New Direction | 54 |
|     4.3.1 Semiconductor Current Relationships | 55 |
|     4.3.2 Static Latchup Criterion: An Incorrect Choice | 58 |

|  |  |
|---|---|
| 4.3.3 Differential Latchup Criterion: A Matter of Stability | 59 |
| 4.3.4 High Level Injection Effects | 63 |
| 4.4 SAFE Space: A Rigorous Definition of the Blocking State | 66 |
| 4.4.1 Symmetric PNPN Structure | 67 |
| 4.4.2 Floating N-well Triode | 69 |
| 4.4.3 Floating Substrate Triode | 71 |
| 4.4.4 General Tetrode | 72 |
| 4.5 Saturation Region Modeling: A New View | 76 |
| 4.5.1 Current Equations | 77 |
| 4.5.2 Differential Resistance | 80 |
| 4.5.3 Holding Current | 81 |
| 4.5.4 Holding Voltage | 83 |
| 4.6 Illustration of Latchup: Type 2 Triggering | 84 |
| 4.6.1 A Sample Analysis | 84 |
| 4.6.2 A SAFE Space Map and Switching Current | 90 |
| 4.7 Modifications to Lumped Element Model: A Useful Perspective | 93 |
| 4.7.1 Transmission Line Model of Latchup | 94 |
| 4.7.2 Transfer Resistance | 99 |
| 4.8 Dynamic Latchup Effects | 104 |
| 4.8.1 Sources of Time Dependence | 105 |
| 4.9 Modeling and Analysis Review | 115 |

**Chapter 5. LATCHUP CHARACTERIZATION** — 117

|  |  |
|---|---|
| 5.1 Measuring Instruments | 118 |
| 5.1.1 Curve Tracer | 118 |
| 5.1.2 Parameter Analyzer | 124 |
| 5.2 Two Terminal Characterization | 126 |
| 5.2.1 Supply Overvoltage Stress | 126 |
| 5.2.2 Supply Overcurrent Stress | 129 |
| 5.3 Three and Four Terminal Characterization | 130 |
| 5.3.1 Voltage Excitation at External Emitter or Base Terminal | 131 |
| 5.3.2 Current Source Excitation on Bypass Resistor Terminal | 133 |
| 5.3.3 Current Source Excitation on External Emitter Terminal | 135 |
| 5.4 Switching Point Characterization | 138 |
| 5.5 Holding Point Characterization | 149 |
| 5.6 Dynamic Triggering | 155 |
| 5.7 Temperature Dependence | 160 |
| 5.8 Non-Electrical Probing | 161 |
| 5.9 Characterization Review | 163 |

**Chapter 6. AVOIDING LATCHUP** — 165

|  |  |
|---|---|
| 6.1 Layout Guidelines | 165 |
| 6.1.1 Guard Structures | 165 |
| 6.1.2 Multiple Well Contacts | 172 |
| 6.1.3 Substrate Contact Ring | 172 |
| 6.1.4 Butted Source Contacts | 173 |
| 6.2 Process Techniques for Bipolar Spoiling | 174 |
| 6.2.1 Lifetime Reduction | 174 |
| 6.2.2 Retarding Base Field | 175 |
| 6.2.3 Schottky Barrier Source/Drain | 178 |
| 6.3 Process Techniques for Bipolar Decoupling | 181 |
| 6.3.1 Epi-CMOS | 181 |
| 6.3.2 Retrograde Well for Lower Sheet Resistance | 184 |

# Contents

|     |     |
| --- | --- |
| 6.3.3 Substrate and Well Bias | 185 |
| 6.3.4 Trench Isolation | 188 |
| 6.4 CMOS Design Considerations | 190 |
| 6.5 Latchup-free Design | 192 |
|     6.5.1 Decouple Vertical Parasitic (DVP) | 193 |
|     6.5.2 Decouple Lateral Parasitic (DLP) | 196 |

**Chapter 7. SUMMARY**     197
7.1 Problem Description     197
7.2 Modeling and Analysis     198
7.3 Characterization     201
7.4 Avoiding Latchup     204

**Appendices**
A. Stability Considerations for PNPN Current-Voltage
    Measurements     209
B. Possible Latchup Characterization Problems     213

**References**     217

**Glossary: Symbol Definitions**     227

**Index**     239

**About the Author**     243

# Figures

**Chapter 2. LATCHUP OVERVIEW**
| | |
|---|---|
| 2-1. Cross section of inverter circuit in N-well CMOS | 8 |
| 2-2. Parasitic bipolar portion of N-well CMOS inverter | 9 |
| 2-3. Complete circuit schematic for N-well CMOS inverter | 11 |
| 2-4. Illustrative PNPN current/voltage characteristics | 12 |
| 2-5. Emitter/base bypass resistances for N-well CMOS inverter | 13 |
| 2-6. Parasitic bipolar portion of 4-terminal PNPN structure | 14 |
| 2-7. Lumped element model for distributed N-well collection | 15 |
| 2-8. Lumped element model for PNPN structure with a floating N-well | 16 |
| 2-9. Latchup criterion for PNPN structure with floating N-well | 18 |
| 2-10. Transistor parameters for simulations of Figs. 2-11 and 2-12 | 19 |
| 2-11. Characteristics of floating well in latched state | 20 |
| 2-12. PN junction biases in latched state | 21 |

**Chapter 3. PROBLEM DESCRIPTION**
| | |
|---|---|
| 3-1. Current path across collector/base junction | 27 |
| 3-2. N-well avalanche triggering of latchup | 27 |
| 3-3. Current path from collector to emitter | 28 |
| 3-4. Punchthrough triggering of latchup | 29 |
| 3-5. Field FET triggering of latchup | 31 |
| 3-6. Triggering taxonomy and latchup sequence | 36 |

**Chapter 4. LATCHUP MODELS AND ANALYSES**
| | |
|---|---|
| 4-1. Point-contact transistor | 39 |
| 4-2. Experimental V/I characteristics for SCR triode | 43 |
| 4-3. Lumped element model for 4-terminal PNPN structure | 47 |
| 4-4. External connections for 4-terminal PNPN device | 55 |
| 4-5. SAFE space representation of the blocking state | 66 |
| 4-6. Illustration of general tetrode latchup condition | 73 |
| 4-7. Parameter values for lumped element model | 85 |
| 4-8. PNPN turn-on characteristics in the blocking state | 86 |
| 4-9. Comparison of static and differential latchup criteria | 87 |
| 4-10. Saturation I/V characteristics for a PNPN structure | 88 |
| 4-11. PNPN saturation bypass and emitter currents vs. terminal current | 89 |
| 4-12. PNPN differential resistance vs. terminal current | 90 |

4-13. A representative map of SAFE space 91
4-14. A comparison of triode and tetrode switching currents 92
4-15. Cross section of parasitic PNPN structure 95
4-16. Transmission line model for parasitic PNPN structure 96
4-17. Transfer resistance definitions 97
4-18. Dependent voltage source model of latchup 98
4-19. Output-excited transfer resistance for high output resistance 99
4-20. Output-excited transfer resistance for low output resistance 101
4-21. Input-excited transfer resistance vs. section lengths 102
4-22. Substrate majority carrier guard structure 103
4-23. Comparison of turn-on time for high and low level injection 106
4-24. Simulated PNPN response to ramped power supply 108
4-25. Blocking state potential distribution 109
4-26. Transient potential distribution 110
4-27. Latched state potential distribution 111
4-28. Electron and hole currents vs. time in nS 112
4-29. Instantaneous behavior 2.9 nS after pulsing LPNP base 113
4-30. Instantaneous behavior 3.4 nS after pulsing LPNP base 114

## Chapter 5. LATCHUP CHARACTERIZATION

5-1. Curve tracer characterization of latchup 119
5-2. Supply overvoltage triggered characteristics 121
5-3. Curve tracer latchup characteristics 122
5-4. Blocking state characterization 125
5-5. Supply overvoltage stress condition 127
5-6. Supply overvoltage latchup characteristics 129
5-7. P-well current triggering 134
5-8. Switching from the blocking state 136
5-9. Points of zero differential resistance 137
5-10. Illustration of 1 mV measurement resolution 139
5-11. Small-signal alpha and beta 141
5-12. Determination of internal substrate resistance 142
5-13. Switching current for PNP-driven triode configuration 143
5-14. Second transistor emitter current at switching point 144
5-15. Determination of internal N-well resistance 145
5-16. Switching current for NPN-driven triode configuration 146
5-17. Second transistor emitter current at the switching point 147
5-18. Comparison of lateral, small-signal betas 148
5-19. Latched state I/V characteristics 150
5-20. Holding point measurement using voltage source 152
5-21. Low impedance state currents 154
5-22. Transient excitation using power supply ramp 155

Figures                                                            xiii

5-23. Critical power supply risetime                               156
5-24. Transient excitation by base injection                       157
5-25. P-well pulse excitation                                      158
5-26. Emitter/base pulse excitation                                159

**Chapter 6. AVOIDING LATCHUP**
6-1. Minority carrier guard in substrate                           168
6-2. P-well PNPN structure                                         169
6-3. Majority carrier guard in well                                170
6-4. Majority carrier guard in substrate                           171
6-5. Schottky barrier source/drain fabrication                     178
6-6. Substrate generator with 2-stage charge pump                  186
6-7. Charge pump waveforms                                         187
6-8. Various lateral isolation technologies                        189
6-9. Worst case latchup consequences                               196

**APPENDICES**
A-1. Circuit schematic for current excitation of PNPN device       210
A-2. Illustration of negative differential resistance measurements 211
B-1. Substrate potential at the switching point                    214

# Preface

## Why a book on latchup?

Latchup has been, and continues to be, a potentially serious CMOS reliability concern. This concern is becoming more widespread with the ascendency of CMOS as the dominant VLSI technology, particularly as parasitic bipolar characteristics continue to improve at ever smaller dimensions on silicon wafers with ever lower defect densities. Although many successful parts have been marketed, latchup solutions have often been ad hoc. Although latchup avoidance techniques have been previously itemized, there has been little quantitative evaluation of prior latchup fixes. What is needed is a more general, more systematic treatment of the latchup problem. Because of the wide variety of CMOS technologies and the long term interest in latchup, some overall guiding principles are needed.

Appreciating the variety of possible triggering mechanisms is key to a real understanding of latchup. This work reviews the origin of each and its effect on the parasitic structure. Each triggering mechanism is classified according to a new taxonomy. In Type 1 triggering one of the parasitic bipolars has been turned on by an external source, and if its collector current is large enough, it turns on the second. In Type 2 triggering the external source causes current to flow through both bypass resistors. If large enough, the bypass current turns on the first transistor, whose collector current then supplements the bypass current in turning on the second. In Type 3 triggering the external source has degraded the blocking state by creating a low impedance path even before the parasitic bipolars are turned on. This taxonomy helps systematize latchup modeling and characterization.

Latchup models in the technical literature are not well developed, and there is little agreement on a specific, yet widely applicable, latchup criterion. Much of this material is, at best, incomplete, and at worst, incorrect. For example, a complete analysis of latchup using the lumped element model is missing from the literature even though the lumped element model for the parasitic PNPN structure was introduced over a decade ago. In addition, much of the theoretical latchup analysis that does appear in the literature rests on an equation for holding current that is conceptually invalid and numerically inaccurate when applied to the holding point. The corrected expression for holding current is provided herein for the first time, as is an expression for holding voltage that fully explains its experimentally observed behavior. However, the present work argues that a switching point description of latchup is easier and more meaningful than a holding point description. A new differential latchup criterion is introduced that precisely describes PNPN switching from the blocking state. It has no fitting parameters, is applicable to any PNPN configuration, and unambiguously specifies the switching current. This differential latchup criterion rigorously defines the switching boundary for the blocking state, and it is illustrated with simple examples.

It has not been clear how to calculate values for the equivalent lumped elements, especially for various resistances in the substrate, where two-dimensional current flow is common. This book demonstrates that substrate resistance is more conveniently viewed as a transfer resistance than as a two terminal element. Transfer resistance values for a wide variety of situations can be found from the lossy transmission line model, which is also useful for understanding voltage and current distributions in the substrate. Transmission line parameters are easily related to the substrate's doping profile and to the relevant layout dimensions.

Latchup characterization has been even less standardized than latchup modeling. There has been little agreement on the correct measurements to be made or on the correct interpretation of the data once it is taken. Characterization techniques evolved from previous work on SCR's, which utilized both junction avalanche and external gating to trigger latchup. When a current source is used for the latter, however, it changes the PNPN configuration from a tetrode to a triode, and the measured switching and holding currents for the latter are not always representative of the

## Why a book on latchup? xvii

former. The switching current can be measured unambiguously for the tetrode using a parameter analyzer with a current source on one of the emitters. Measurements made this way agree very closely with the new differential latchup criterion.

Because the parasitic bipolars in the PNPN structure each have a finite response time, switching from the blocking state requires a stronger excitation when that excitation is applied for only a short duration. Understanding this behavior is important for quantitatively describing triggering conditions encountered during actual operation. The key is knowing whether the transient excitation can shift the dynamic operating point to the switching border of SAFE space.

Latchup is avoided by operating CMOS circuits solely in SAFE space. This strategy can be implemented using layout guidelines, processing techniques, or some combination of both. Previous implementations are critiqued from a fresh perspective that enables the designer to judge their applicability. Guard structures are explained, and their various forms are compared.

Processing techniques prevent latchup by either spoiling the parasitic bipolar characteristics or decoupling the parasitic NPN and PNP from one another. Although both have met with some success, the latter is more compatible with today's CMOS technologies and has been more widely implemented. Specific decoupling recommendations are made and discussed relative to general CMOS design considerations.

If proper attention is given to these process and layout guidelines, CMOS will be free of latchup, even for micron and sub-micron lithography. This book discusses how these guidelines can be used to restrict all parasitic PNPN devices to SAFE space.

# Acknowledgements

Many people have contributed to the writing of this book through their papers and presentations on latchup. I have carefully referenced specific contributions from the technical literature throughout the pages, but the bibliography is not intended to be totally exhaustive (and hopefully not totally exhausting). I have enjoyed many valuable and stimulating discussions with colleagues working to eliminate the latchup problem, and their generosity in providing information has improved the quality of this work. Of the many I would like to specifically cite H. Zappe and I. Leventhal, whose theses have enhanced the latchup literature, as well as W. Craig, M. Hargrove, G. Hu, W. Lynch, D. Nelsen, A. Ochoa, R. Rung, and C. Sodini.

Thanks also goes to P. Chatterjee for an invitation to review recent latchup developments at the 1983 Device Research Conference. Preparing this talk planted the seed for the present monograph. Its growth was nourished by material originally prepared for several latchup seminars given in 1984 and, at the invitation of Prof. Viswanathan and J. Chen, for a UCLA short course given in June of that year and again in Jan. 1985. An encapsulated review of this material was presented at the 1984 IEDM, thanks to R. Davies' invitation.

I am also very grateful to the IBM Corporation in general, and to W. T. Siegle and my manager J. Hiltebeitel in particular, for making possible a year's sabbatical at MIT during which the major portion of this monograph came to fruition. A special note of thanks is due C. Tillman and R. Reynolds of the IBM Cambridge Scientific Center for their patient and able assistance in using QPRINT and transferring data files. Thanks also go to B. El-Kareh and R. Bombard for timely advice and assistance in the final preparation of this book.

# LATCHUP IN CMOS TECHNOLOGY

# Chapter 1

## INTRODUCTION

Latchup is a key concern to bulk CMOS. It stems from parasitic bipolar transistors, which are structurally inherent to bulk CMOS. These transistors can be activated in various ways, and as CMOS technologies are scaled down, both in dimension and in circuit delay, this variety grows. Under certain conditions the activated transistors can dominate circuit behavior. However, with proper process and layout design, CMOS chips can be operated under relatively harsh conditions without ever encountering latchup. What is proper can vary with application - the type of chip, its performance, package, allowable cost, etc. This book will provide the insight - as well as some useful analysis, characterization, and avoidance techniques - for assessing what is proper in each application.

This book views latchup within the larger context of CMOS technology and design. Accordingly, it begins with a review of why CMOS is emerging as the most important Very Large Scale Integration (VLSI) technology, then provides a historical perspective on latchup prevention. This introductory chapter ends with a discussion of why latchup deserves special attention in VLSI CMOS.

### 1.1 CMOS: The Emerging VLSI Technology

Although CMOS was first proposed in 1962 [Sah-62], its early development was primarily limited to particular applications. One such application was battery operated wrist watches and calculators, where low power and excellent noise immunity are of paramount importance. Another was satellite and missile circuits, where insensitivity to moderately high radiation levels is important. One type of CMOS technology was particularly suited to the latter. In CMOS-on-sapphire the N- and P-channel FET's are dielectrically isolated from one another and consequently cannot experience latchup. However, because it is fabricated on an expensive substrate and suffers from poor electrical leakage properties, CMOS-on-sapphire is not widely used. Throughout this book the term "bulk CMOS" is used to denote CMOS fabricated strictly on silicon, either in an epitaxial layer grown on the substrate (epi-CMOS) or directly in the substrate.

Because performance and density were not dominant considerations in these applications, CMOS lithography and fabrication techniques generally lagged behind those of NMOS and bipolar technologies. In recent years, however, CMOS has become an important semiconductor technology, and there are many indications that by 1990 half the integrated circuit market will be served by CMOS technologies [Davies-83].

This change is occuring for two fundamental reasons. (1) Further development of bipolar and NMOS technologies for many applications tends to be limited by power dissipation. (2) Increased NMOS processing complexity and clever CMOS processing techniques, such as self-aligned field regions and maskless source/drain implants, have nearly eliminated the differential in fabrication costs between NMOS and CMOS. In addition, unique CMOS attributes have made it especially attractive to a wide variety of chip designers.

Long recognized for its low power advantage, CMOS allows more circuits within the power constraints of a given package. Compared to bipolar or even NMOS, the greater number of circuits permitted within the package and the resulting reduction in package interfaces can be used by system architects to achieve better system performance. CMOS can also reduce system cost by eliminating the need for cooling fans and reducing the size of power supplies.

Another important CMOS advantage is ratioless logic design. Coupled with full logic swings between ground and the power supply voltage, it gives CMOS a decided edge in choosing power supply voltage. Larger than standard voltages can be applied to

Introduction 3

achieve higher performance, or lower voltage can be applied to take full advantage of increased circuit densities resulting from improved lithographic and fabrication capabilities in scaled CMOS technologies. Full voltage swings also lessen the need for bootstrapping both off-chip drivers and on-chip drivers (such as word line drivers on memory chips).

Still another advantage is the noise immunity provided by the relatively low impedance path of the "on" device in any logic gate and by the reverse biased well. The latter is especially important to memory chip designers because of the extra protection against alpha particles. Some protection of this sort is necessary for the low capacitance storage nodes associated with high density memory cells. This noise immunity also makes it possible to reduce power supply voltage as dimensions are scaled down.

Performance in CMOS technologies is less sensitive than in NMOS to variations in device parameters caused by processing tolerances, a feature particularly attractive to operational amplifiers. Analog designs in CMOS also benefit from a bilateral switch (consisting of parallel N- and P-channel FET's) free from voltage loss caused by the body effect on FET threshold. Because of its versatility, CMOS is a natural choice for both analog and digital circuits, and some applications combine the need for both on one chip.

## 1.2 Latchup Prevention: A Historical Perspective

During the same years that CMOS was assuming an important role in VLSI technology, several techniques for controlling latchup were being investigated. Early results are reviewed in [Estreich-80], and recent results in [Troutman-84b]. Latchup has always been a potentially serious problem for bulk CMOS, and the trend to ever smaller device dimensions exacerbates the problem.

Latchup in bulk CMOS stems from the PNPN structure formed by a parasitic NPN and a parasitic PNP bipolar transistor. One of these bipolar transistors is a vertical device formed in the well, the other a lateral transistor formed in the substrate. The two are inherently connected as a pair, with the reverse biased well-to-substrate junction acting as the collector for both transistors.

Latchup can be prevented by holding below unity the sum of effective small-signal, common base current gains for the two transistors. Consequently, there are two general strategies for avoiding latchup - bipolar spoiling and bipolar decoupling. All latchup protection techniques follow at least one of these two strategies. In the first, one deliberately includes steps in CMOS fabrication to spoil transistor action by reducing either carrier transport or injection. Examples of spoiling techniques include gold doping, neutron irradiation, a base retarding field, and Schottky barrier source/drains. In the second, one concentrates on decoupling the bipolar transistors to prevent one transistor from turning on another, using either layout or processing decoupling techniques. Layout examples include butted contacts and both majority and minority carrier guard rings. Process examples include using a lightly doped epitaxial layer on a highly doped substrate, retrograde wells, and trench isolation.

The decoupling strategy for controlling latchup arises from the realization that years of development have provided the integrated circuit industry silicon wafers with long minority carrier lifetimes, and that many products, such as DRAM's, are heavily dependent on these high quality wafers. In addition, shrinking lateral dimensions, the sine qua non of integrated circuit manufacturing, generally reduces basewidths. Trying to spoil the bipolars is tantamount to fighting both mother nature and industrial trends. That is why most recent efforts to control latchup are more realistically described as bipolar decoupling. Indeed, there is growing recognition that such accidently good bipolar devices should be added to the chip designer's tool kit once their decoupling is perfected.

Currently, almost all bulk CMOS technologies using 2 micron (and below) lithography are fabricated in a lightly doped epitaxial layer grown on a highly doped substrate (epi-CMOS). They all use some form of guard rings to avoid latchup, although rules for using guard rings are far from universal. A few reported technologies use a retrograde well. Almost without exception, the

Introduction                                                                                     5

modern thrust in latchup control is to decouple the parasitic bipolars.

## 1.3  Latchup in VLSI CMOS: A Continuing Concern

Although today's CMOS technologies provide some degree of latchup protection, future CMOS technologies will find the problem more difficult. Smaller geometries will mean decreased base widths for the parasitic bipolars, which will necessitate better decoupling. Smaller spacings will also tax device isolation and could increase the likelihood of punchthrough and field FET triggering, as well as lateral bipolar action in the well. Low doping levels in the well and in the epitaxial layer, advantageous for low source/drain capacitance, aggravate these problems.

Minimum $N^+/P^+$ spacing is very important to some CMOS circuits, such as six device static RAM cells. Careful attention to doping profiles at the lateral well boundary is one possible course for future isolation. Shallow wells waste less space because of reduced lateral diffusion, and self-aligning field implants to the well edge can hold this wasted space to a minimum. Trench isolation is another possible course. Trench depths equal to or greater than the well depth prevent current flow to and from the well at its lateral boundary. Each of these courses can increase process complexity, and deciding whether they are needed for latchup control requires a detailed understanding of alternative solutions and a careful evaluation of the tradeoffs.

In addition, as CMOS permeates a wider range of product designs, problems associated with fast switching at the I/O pads will become a concern for more chip designers. They will need to pay greater attention to either avoiding transient over and undershoots or following prescribed layout procedures to prevent such transients from causing latchup. This problem is especially acute in CMOS because current flows only during switching transitions, causing inductive voltages during both polarity changes of each clock cycle.

The obvious concern in VLSI CMOS is the effect of small dimensions on device characteristics. For example, smaller basewidth means higher bipolar gains, with the biggest increase expected for the lateral device. Smaller dimensions also increase the likelihood of punchthrough and parasitic FET triggering. Less obvious, perhaps, are the better bipolar characteristics resulting from the high minority carrier lifetimes necessary to dynamic memories, which tend to drive VLSI technology.

An important final note, sometimes overlooked, is that whatever latchup solutions are adopted, they should apply to all CMOS designs - digital and analog; dynamic, static, and programmable memories; gate array, standard cell, and custom logic. These solutions should be compatible with large volume manufacturing and not add significantly to manufacturing cost. Ideally, they should permit the minimum $N^+/P^+$ spacing allowed lithographically.

Latchup is a challenging problem to VLSI CMOS. Subsequent chapters will elucidate the latchup problem, discuss how it is characterized and modeled, and, most importantly, describe how it is avoided. Curing the latchup problem is this book's goal.

# Chapter 2

## LATCHUP OVERVIEW

CMOS circuits cannot accidentally latch if they operate solely in SAFE space. Before beginning a survey to delineate the boundaries of this arcane territory and to explore its interior and neighboring regions, we shall take a brief excursion to familiarize ourselves with the terrain.

This chapter describes the structural origin of latchup and the lumped element model commonly used to analyze its behavior. Using a PNPN structure with a floating N-well as a specific example, it develops a latchup criterion and follows the latchup sequence across state boundaries.

### 2.1 Structural Origins and a Lumped Element Model

Latchup lurks in the very structure required to fabricate bulk CMOS. To have both N- and P-channel FETs, it is necessary to have both P and N type background material. This is commonly accomplished by starting with a silicon wafer of one type (say, P type, as in Figure 2-1) and creating in it regions of the opposite type. Early in the CMOS processing sequence, areas of the wafer are exposed to receive (by diffusion or by ion implantation) a dopant species opposite to that in the starting wafer. The concentration of this dopant species is carefully controlled to counter-dope selected areas to a prescribed depth on the order of 2 to 5 microns. The resulting volumes, where the silicon substrate has been counterdoped, are known as wells or tubs. P-channel FETs are built in the N-well of Figure 2-1, while N-channel devices are built directly in the P type substrate. Unfortunately, the FETs are not the only structures fabricated; PNPN devices consisting of parasitic bipolar transistors are also created.

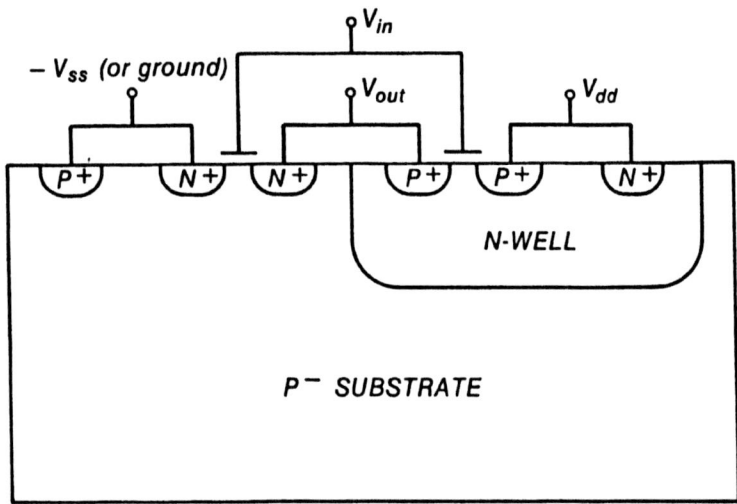

Figure 2-1. Cross section of inverter circuit in N-well CMOS.

First, there are vertical PNP bipolar transistors formed by the $P^+$ source/drain diffusion, the N-well, and the P-type substrate. When forward biased, any $P^+$ diffusion can serve as an emitter and inject holes into the N-well base. The reverse biased junction formed by the N-well and substrate then collects the unrecombined holes. Secondly, there are lateral NPN bipolar transistors formed by any $N^+$ source/drain diffusion, the P-type substrate, and the N-well. In this case electrons injected from a $N^+$ diffusion into the substrate can be collected by the reverse biased N-well.

Figure 2-2 superposes the bipolar equivalent circuit for the inverter on its cross section. There are two vertical PNP and two lateral NPN transistors. Both the N-well and the P-type substrate serve two functions. The N-well is the base for either vertical PNP and the collector for either lateral NPN. Likewise, the P-type substrate serves as the base for either lateral NPN and the collector for either vertical PNP. Each of the collector regions can develop a voltage drop between the collector/base junction and the collector contact, which can be modelled as a collector resistance. In addition, current flowing through some fraction of the collector resistance can forward bias the emitter/base junction

# Latchup Overview

of the opposite bipolar device if the IR drop exceeds several tenths of a volt.

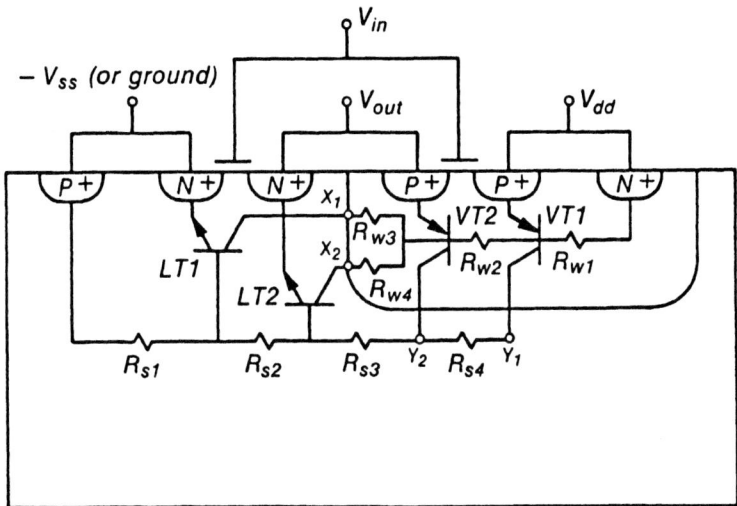

Figure 2-2. Parasitic bipolar portion of N-well CMOS inverter.

The resistor $R_{w1}$ in Figure 2-2 represents resistance from the N-well contact to the intrinsic base region of the first vertical PNP while the resistor $R_{w2}$ represents resistance from the intrinsic base region of the first to the second. $R_{w3}(R_{w4})$ represents the resistance from the intrinsic base region of VT2 to the point $X_1(X_2)$ where the current from LT1(LT2) is collected. Electrons injected from the $N^+$ emitters are not all collected at one point, but we simplify the picture somewhat to nurture the lumped equivalent model. In some cases (particularly for epi-CMOS) the points $X_1$ and $X_2$ can be considered the same; then the parallel combination of $R_{w3}$ and $R_{w4}$ can be replaced by a single resistor. In a similar manner, $R_{s1}$ represents resistance from the substrate contact to the intrinsic base region of the first lateral NPN, and $R_{s2}$ represents resistance between the base regions of LT1 and LT2. The resistor $R_{s3}(R_{s4})$ represents resistance from the base intrinsic region of LT2 to the point $Y_1(Y_2)$ where the current from VT1(VT2) is collected. Again we simplify the physical behavior for the sake of a lumped equivalent circuit. The points $Y_1$ and $Y_2$ are generally easier to locate than $X_1$ and $X_2$ because base current flow is better defined in the vertical bipolar.

Details of the inverter cross section shown in Figure 2-1 can vary without changing the basic considerations under discussion. For example, contact to the P-type substrate is often made via the entire backside of the wafer rather than through a topside $P^+$ diffusion. The lumped equivalent circuit shown in Figure 2-2 is still valid (although resistor values may no longer be the same), but the accessible end of resistor $R_{s1}$ then connects to the backside contact. Also, most CMOS technologies support a butted contact, where an $N^+$ and a $P^+$ diffusion actually touch and are contacted directly with aluminum across the metallurgical junction formed by the two diffusions. Such a contact could replace the combination of the $N^+$ diffusion forming the contact to the N-well and the nearby $P^+$ source, which is shown externally shorted to the $N^+$ diffusion in Figure 2-1 and in Figure 2-2. It could also replace the topside $P^+$ substrate contact and a nearby $N^+$ source. Butted contacts can be used, of course, only for diffusions that would be shorted together anyway and would not be appropriate if two or more P-channel or N-channel devices were to be placed in series. To preserve generality, we shall continue using the cross section depicted in Figure 2-1.

The total equivalent circuit for the inverter circuit has the form shown in Figure 2-3. The simple CMOS inverter, consisting of an N-channel driver and a P-channel load, is the desired result of a given layout. In parallel with it, however, is the circuit formed by the parasitic bipolars and associated resistors. Under normal operation the circuit performs as an inverter, and the bipolar portion can be ignored. Under certain conditions, which are explored later, the operation of the bipolars can dominate the behavior of the total circuit. In particular, if the bipolar circuit switches from its normally high impedance state into its low impedance state, the power supply then sees a low impedance path to ground. If the current from the supply is not limited somehow, there can be an irreversible change, such as the fusing of an aluminum line somewhere in the chip's power supply or ground line. Even if the current is limited so that no irreversible change occurs, however, the PNPN's low impedance state could cause a circuit to malfunction. To guarantee accurate circuit behavior, the PNPN must remain in its high impedance state.

# Latchup Overview

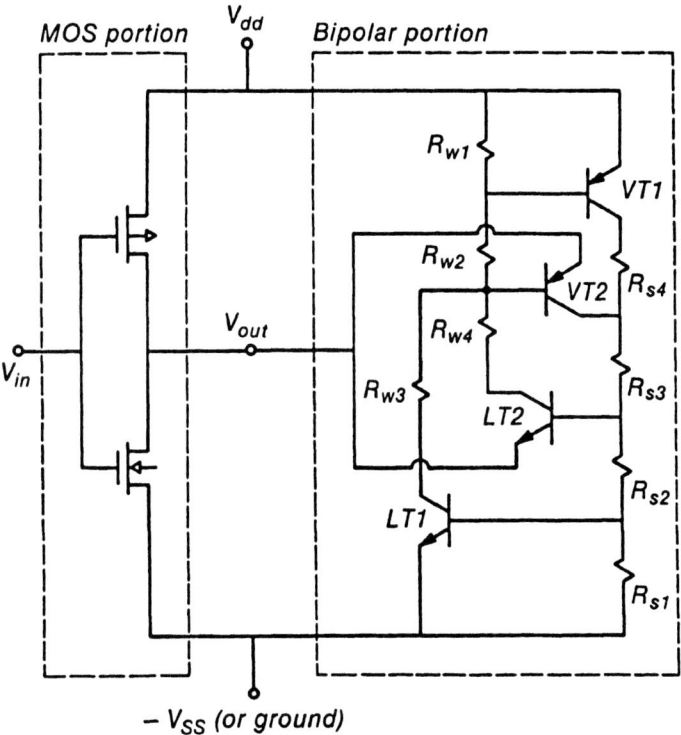

Figure 2-3. Complete circuit schematic for N-well CMOS inverter.

The current/voltage characteristic shown in Figure 2-4 schematically depicts the high and low impedance states. Any characteristic observed for a specific PNPN structure depends on how switching is initiated. Since a PNPN structure is really a 4-terminal device, the observed characteristic also depends on the terminal conditions of the N-well and substrate contacts. Different connections for these terminals could produce even qualitatively different characteristics. In Figure 2-4 these are shown short-circuited. Two key points are identified on this characteristic - $(V_s, I_s)$ and $(V_h, I_h)$. The terminal current $I_s$ marks the transition from the high impedance region (also referred to as the blocking state or OFF region) to the negative differential resistance region, and $I_h$ marks the transition from the negative differential resistance region to the low impedance region (also referred to as the latched or ON region). The blocking and latched regions are also referred to as states since the parasitic

PNPN device can stably reside in each and can be switched from one to the other.

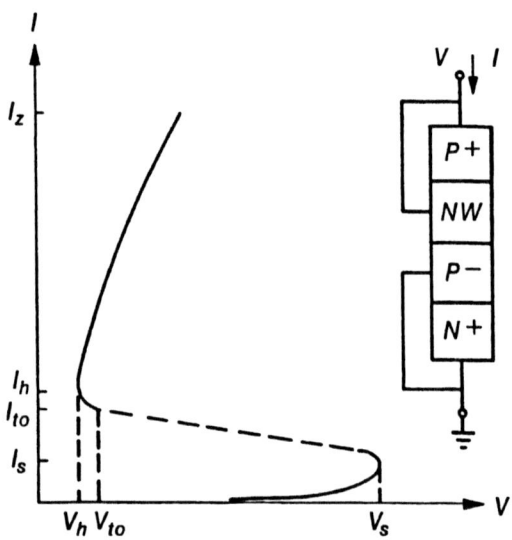

Figure 2-4.  Illustrative PNPN current/voltage characteristics. A stable high impedance state is defined by $I < I_s$, and a low impedance state by $I_h < I < I_s$. At the turn-off point reverse bias on the center junction is zero.

Latchup terminology is far from standard. In the literature $I_s$ is sometimes referred to as switching current and sometimes as critical current, and $I_h$ is usually referred to as holding current. When it is large enough to cause switching from the blocking to the latched state, the current flowing at any terminal of interest is sometimes said to "trigger" latchup and as such is referred to as trigger current, but such a notable current is also referred to as threshold current. Differences in terminology stem partly from different terminal conditions during measurement and partly from incorrect simplifying assumptions made in previous latchup modeling work. The chapters on modeling and characterization examine these differences in more detail.

When the PNPN switches from the blocking to the latched state, reverse bias on the center junction decreases to zero at a point in the differential negative resistance region often called the "turn-off" point. This is not a stable point, and increasing device

Latchup Overview 13

current then causes the center junction to forward bias. At the holding point this junction is forward biased by an amount $V_r = V_{to} - V_h$.

An inverter's latchup behavior depends on the terminal voltage $V_{out}$. If $V_{out} = V_{dd}$, the transistor LT2 is removed from consideration since there is no way current flowing through the resistors in Figure 2-2 can turn it on. The remaining three transistors can be turned on, however, if current through the appropriate emitter/base bypass resistors produces a sufficiently large base/emitter voltage. Figure 2-5 lists the appropriate bypass resistors for the two static conditions $V_{out} = V_{dd}$ and $V_{out} = -V_{ss}$. During switching of the inverter, $V_{out}$ is changing from one of these values to the other, and there is some time period when both LT2 and VT2 could possibly be turned on, but such an event is highly unlikely.

| Transistor | Output high | Output low |
|---|---|---|
| VT1 | $R_{w1}$ | $R_{w1}$ |
| VT2 | $R_{w1} + R_{w2}$ | OFF |
| LT1 | $R_{s1}$ | $R_{s1}$ |
| LT2 | OFF | $R_{s1} + R_{s2}$ |

Figure 2-5. Emitter/base bypass resistances for N-well CMOS inverter.

An important distinction exists between VT1 and VT2 when $V_{out} = V_{dd}$. Although both transistors can be turned on by a sufficiently large bypass current, emitter current into VT2 is limited by the P-channel load FET, assuming other connections to the output node cannot supply current. The emitter of VT1 is connected directly to the power supply. Likewise, LT2 emitter current is limited by the N-channel driver device when $V_{out} = -V_{ss}$. Thus, the combination of VT1 and LT1 is more likely to sustain latchup once it occurs.

On an actual CMOS chip there can be many combinations of parasitic vertical and lateral bipolar transistors that potentially participate in latchup, some of which can form circuits more complicated than an inverter. This circuit complexity tends to obscure the device physics behind latchup behavior. The inverter itself is more complicated than is necessary to describe fundamental latchup effects. Consequently, studies often utilize the elementary 4-terminal PNPN structure shown in Figure 2-6, with separate, external access to all four terminals, to measure and model latchup behavior. It consists of only one vertical and one lateral parasitic transistor. Although it is the simplest structure that can exhibit latchup, it represents the most latchup sensitive part of any CMOS circuit, and much of the latchup data reported has been measured on 4-terminal PNPN devices. The corresponding lumped element circuit model is also shown superposed in Figure 2-6.

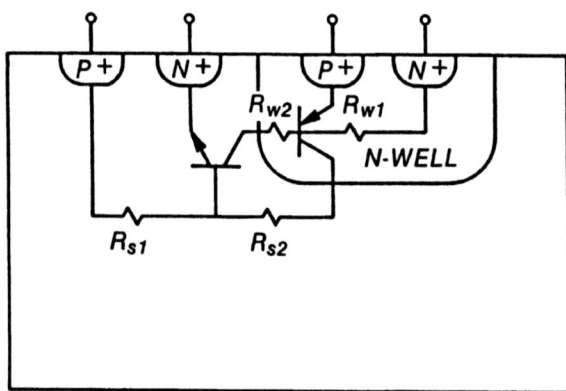

Figure 2-6. Parasitic bipolar portion of 4-terminal PNPN structure. This is the most common version of the lumped element model.

An additional feature has been shown to explain more accurately latchup data for non-epitaxial CMOS [Leventhal-84]. In epitaxial CMOS, lateral NPN base current is collected mostly by the N-well sidewall junction (especially the sidewall facing the $N^+$ emitter) because the high/low junction at the substrate/epi transition prevents electrons from diffusing deep into the substrate. In non-epi CMOS, however, a sizeable fraction of this base current is collected by the junction located at the N-well bottom. Thus, some fraction of the lateral NPN collector current does not flow through $R_{w1}$ and does not forward bias the vertical PNP emitter/base junction. To account for this non-biasing path, the well resistor $R_{w3}$ has been added in the schematic of

Figure 2-7. A similar schematic has also been used to model the PNPN structure when the $N^+$ and $P^+$ diffusions in the well are reversed [Fang-84]. In that case the ohmic drop created by current flowing through the well junction to the well contact can be significantly less than the drop created if the same current flowed under the $P^+$ diffusion to a well contact on the other side.

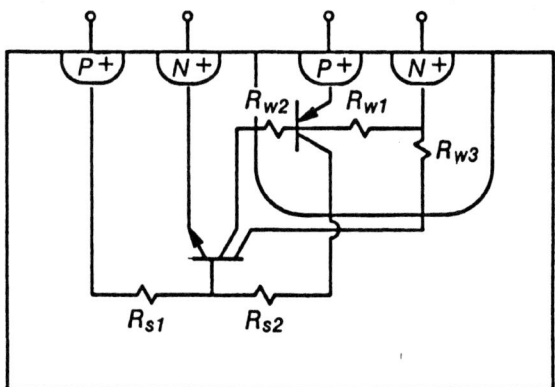

Figure 2-7. Lumped element model for distributed N-well collection. This version accounts for distributed collection of lateral base current by the N-well.

Even though the lumped element models of the previous discussion were presented for an N-well technology, they are equally valid for P-well or twin well. The only differences are the resistor and transistor labels. Labels $R_{s1}$ and $R_{w1}$, as well as $R_{s2}$ and $R_{w2}$, should then be interchanged to delineate which are substrate resistors and which are well resistors (i.e., which are isolated from the substrate). The necessary information for the transistors is conveyed by denoting whether the NPN is a lateral device (LNPN) or a vertical device (VNPN) and by similarly labeling the PNP.

## 2.2 An Example: Latchup in a Floating N-well Structure

Let us now examine switching from the high to the low impedance state in more detail. As a concrete example consider the 4-terminal PNPN structure shown in Figure 2-6 with external resistance $R_x$ in series with the power supply. For simplification the $R_{w1}$ terminal is left unconnected, which is equivalent to floating the N-well. Such a condition can actually occur if a small N-well contact fails to open during the etching step.

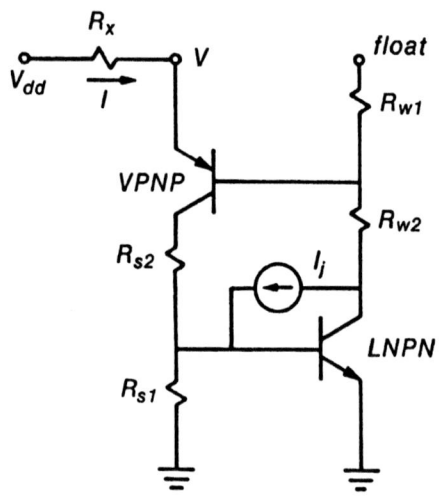

Figure 2-8. Lumped element model for PNPN structure with a floating N-well.

The corresponding lumped element circuit is shown in Figure 2-8. Before either bipolar transistor is turned on, the voltage $V_{dd}$ is dropped entirely across the N-well to substrate junction, reverse biasing the collector/base junctions for both the lateral and vertical bipolars. Let us assume that raising $V_{dd}$ causes the collector/base junction current $I_j$ to begin increasing due to surface avalanche breakdown between the N-well and the highly doped substrate field region. The PNP transistor turns on, and its base current is given by $I_{bp} = I_j + I_{cn}$. The resulting collector current is $I_{cp} = \beta_p I_{bp}$, and the (initially negligible) NPN collector current is $I_{cn} = \beta_n I_{bn}$. The collector/base junction voltage is reduced slightly from $V_{dd}$ by the voltage drop across resistors $R_x$, $R_{s1}$, and $R_{s2}$. In general $\beta_p$ is a function of $I_{bp}$ and may be small if $I_j$ is small.

Suppose now that $V_{dd}$ is raised so that $I_j$ increases still more. Eventually the voltage drop across $R_{s1}$ is large enough to raise $I_{cn}$ well above $I_j$. Then a small change in base current $\Delta I_{bn}$ produces a change in PNP collector current, and might, in some unusual case, produce a change in junction current $\Delta I_j$. The change in collector current also depends on any change in junction current. The general result is $\Delta I_{cp} = \beta_{ps}(\Delta I_j + \beta_{ns}\Delta I_{bn})$, where $\beta_{ns}$ and $\beta_{ps}$ are the NPN and PNP small-signal common emitter current gains, respectively. The resulting current change into the base is then $\Delta I_{bn}' = \kappa(\Delta I_j + \Delta I_{cp})$, where $\kappa = G_{bn}/(G_{bn} + 1/R_{s1})$ is the ratio of small-signal conductance $G_{bn}$ looking into the NPN base divided by the total parallel conductance at the base node. For small $I_j$ values the voltage drop across $R_{s1}$ is very small, and the NPN is turned on only very slightly so that $\kappa$ is very nearly zero. The conductance $G_{bn}$ rises exponentially with current through $R_{s1}$, and $\kappa$ approaches unity for $G_{bn} >> 1/R_{s1}$. Finally, the total loop gain for the PNPN circuit is given in general by

$$\frac{\Delta I_{bn}'}{\Delta I_{bn}} = \kappa \left[ (1 + \beta_{ps}) \frac{\Delta I_j}{\Delta I_{bn}} + \beta_{ns}\beta_{ps} \right]. \tag{2-1}$$

The circuit becomes unstable when the positive loop gain reaches or exceeds unity, i. e., when

$$\beta_{ns}\beta_{ps} + (1 + \beta_{ps}) \frac{\Delta I_j}{\Delta I_{bn}} \geq \frac{1}{\kappa}. \tag{2-2}$$

For the case of a PNPN diode, $R_{s1}$ is absent, and its value is infinity. Then $\kappa$ is equal to one, and a small beta product (unity if $\Delta I_j = 0$) is sufficient to cause latchup. For the case of a PNPN triode, however, $R_{s1}$ is finite, and the right-hand side of the above equation is much larger than one, necessitating a much larger beta product before latchup can occur.

The term involving $\Delta I_j$ in the above latchup criterion can be neglected when latchup is triggered by current flow through $R_{s1}$ rather than by avalanche breakdown of the center junction. Increasing NPN emitter current causes $\kappa$ to increase monotonically and $\beta_{ns}$ to first increase and then, because of high level injection effects, to decrease. Similarly, increasing PNP emitter current causes $\beta_{ps}$ to first increase then decrease. Thus, the left-hand side of (2-2) first increases, then decreases with increasing NPN emitter current while the right-hand side decreases monotonically. The first intersection of the two curves for increasing current determines the switching current, as illustrated in Figure 2-9. A good approximation for the switching current is to set the left hand side equal to its peak value to find the critical value for $\kappa$, then use the appropriate transistor model to solve for the NPN emitter and bypass currents.[1] By making $R_{s1}$ small enough, it is possible to guarantee the curves do not cross, and that latchup is not possible.

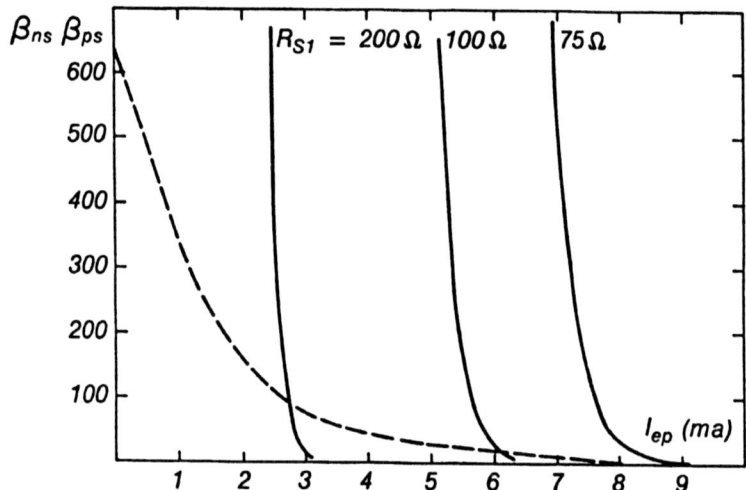

Figure 2-9. Latchup criterion for PNPN structure with floating N-well. Dashed line is beta product, solid lines are $1/\kappa$ for various $R_{s1}$ values. Latchup occurs when $I_{ep}$ reaches the level defined by the intersection of dashed and solid lines.

When the latchup condition does hold, the PNPN structure can be triggered to switch into the low impedance state. The switching current is approximately $I_s = V_{ben,s}/R_{s1}$, where $V_{ben,s}$ is the base emitter bias required to turn on the NPN sufficiently to achieve the critical value of $\kappa$ described above, and the corresponding terminal voltage, or switching voltage, is $V_s = V_{dd} - R_x I_s$. The

---

[1] Explicit equations for the switching current and critical forward bias are derived in Chapter 4.

switching voltage depends on external series resistance (via the "load line"), and switching current and voltage both depend on bypass resistance, but neither depends on the interconnection resistors $R_{s2}$ and $R_{w2}$.

| Parameter | NPN | PNP |
|---|---|---|
| Emitter/base saturation current $I_s$ | 10 pA | 10 pA |
| Collector/base saturation current $I_r$ | 1 nA | 1 nA |
| Forward alpha $\alpha_f$ | 0.90 | 0.95 |
| Reverse alpha $\alpha_r$ | 0.009 | 0.0095 |
| Thermal voltage $V_t$ | 0.026 V | 0.026 V |

Figure 2-10. Transistor parameters for simulations of Figs. 2-11 and 2-12.

After switching, both collector currents rise until the transistors saturate, and now the characteristics do depend on the values of the interconnection resistors. To illustrate, let us assume a CMOS process for which $R_{s2}$ is negligible. In general, however, $R_{w2}$ increases with P+ emitter distance from the collecting sidewall of the N-well. As a further simplification, our numerical example uses Ebers-Moll transistor models with constant current gain and the parameter values listed in Figure 2-10. Figure 2-11 reveals that increasing $R_{w2}$ causes holding current to decrease slightly, holding voltage to increase, and the differential resistance of the latched state to increase nearly linearly with $R_{w2}$. In addition, at higher current the PNPN device switches back to the blocking state, so that the latched state exists only over a range of terminal current. This range shrinks for larger $R_{w2}$.

This behavior can be understood by using Kirchoff's voltage equation for the interconnection network, which says that the voltage drop across $R_{w2}$ equals the difference in reverse drives between the NPN and PNP, to express PNP reverse current $I_{rp}$ as a function of terminal current I, i.e.,

$$I_{rp} = \frac{[(V_{bcn} - V_{cbp}) - (1 - \alpha_{fp})IR_{w2}]}{(1 - \alpha_{fp}\alpha_{rp})R_{w2}}, \qquad (2-3)$$

where $V_{bcn}$ is the reverse drive on the NPN (base/collector forward bias), $V_{cbp}$ is the reverse drive on the PNP, and $\alpha_{fp}(\alpha_{rp})$ is the PNP forward (reverse) alpha. Forward drive on both transistors increases with increasing PNPN device current (when forced by an external current source, for example), which in turn causes the NPN reverse drive to increase and the PNP reverse drive to decrease as shown in Figure 2-12. Eventually the reverse drive on the PNP disappears altogether, and its collector/base becomes reverse biased. The current for which the PNP leaves saturation decreases at larger values of $R_{w2}$.

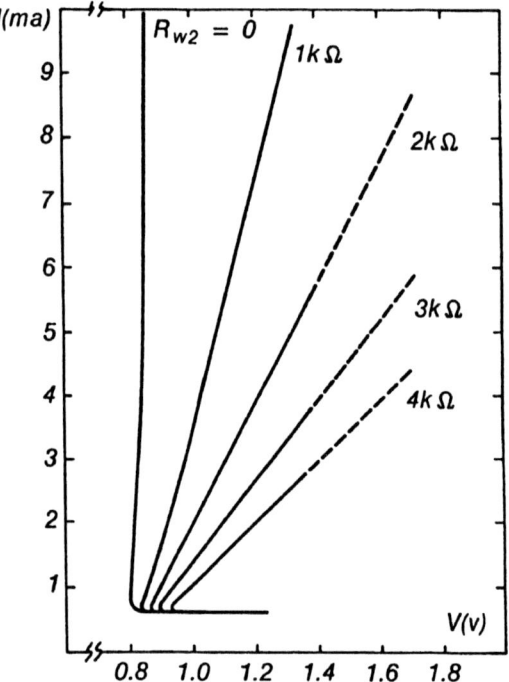

Figure 2-11. Characteristics of floating well in latched state. Solid curves denote both transistors are saturated; dashed, only NPN is saturated. Switching from latched to blocking state occurs at approximately V = 1.2 and 1.7 volts. ($R_{s1}$ = 1kΩ; $R_{s2}$ = 0.)

# Latchup Overview

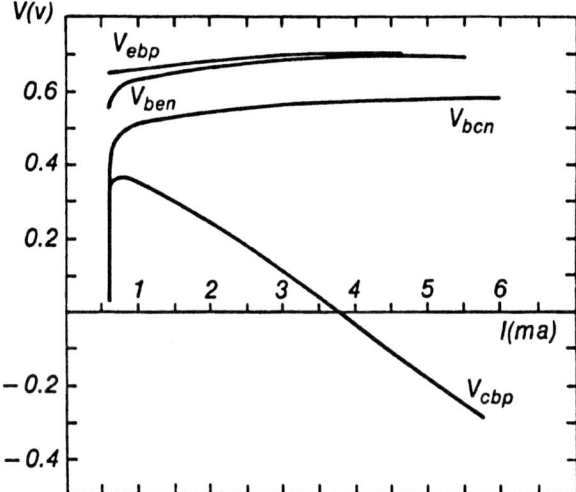

Figure 2-12. PN junction biases in latched state. The PNP is in the forward active region for I ≥ 3.75mA. ($R_{s1} = 1$kΩ; $R_{s2} = 0$; $R_{w2} = 3$kΩ.)

Discussion of latchup in this example of a floating N-well triode focused on the key aspects effecting switching. Switching from the high impedance state occurs when sufficient loop gain exists, and switching from the low impedance state occurs when terminal current is brought below the holding current or when it exceeds some critical value that depends on the interconnection resistance. In the low impedance state at least one transistor is in saturation, which for the case just considered was the lateral NPN. To keep the example simple, we have made some approximations that are not strictly true in the more general tetrode case. Destabilization of the blocking state will be re-examined in Chapter 4, where PNPN current/voltage characteristics are also scrutinized in far greater detail.

# Chapter 3

## PROBLEM DESCRIPTION

We now begin preparing for our systematic survey of SAFE space. Recurring features can help us plan our approach.

This chapter begins by describing latchup as switching from the PNPN's blocking state to its latched state, and the conditions necessary for this switching are precisely stated in general terms. It then investigates all possible causes for this switching and organizes them into three basic categories. Members of a category all share a fundamental latchup sequence.

### 3.1 Blocking and Latched States

The parasitic latch circuit is formed by at least one parasitic PNP and at least one parasitic NPN bipolar transistor. Normally these transistors are off, and the latch is said to be in the blocking state. In this state the latch presents a very high impedance between the power supply and ground.

Under certain conditions the latch can be triggered into a low impedance state (also referred to as the latched state), and if the latch remains in the low impedance state after the excitation is removed, latchup is described as self-sustaining. If the latch returns to its high impedance state after the removal of the excitation, latchup is only temporary and is not self-sustaining.

When latchup occurs, the result is either a momentary or a permanent loss of circuit function. Since the latch has switched into a low impedance state, large currents are possible. If there is no current limiting mechanism (such as a resistor in series with the latch), the resulting high current can cause an aluminum line to evaporate. Even temporary latchup can result in permanent damage if current is not limited.

## 3.2 Conditions Necessary for Latch-up

Several conditions must obtain before the bipolar portion of the circuit can latch (switch into its low impedance state).

1. The loop gain of the relevant PNPN configuration must exceed unity for the circuit to switch. This loop gain is illustrated in section 2.2 and is rigorously defined in Chapter 4, where the condition for latchup is presented in terms of effective injection efficiencies and small-signal current gains for the parasitic bipolar transistors.

2. A bias condition must exist such that both bipolars are turned on long enough for current through the blocking junction to reach the level defined as switching current. Turn-on is usually caused by externally excited current flow through one or both emitter/base bypass resistors. Common external current excitations are discussed in the next section.

3. The bias supply and associated circuits must be capable of supplying current at least equal to the switching current for the PNPN structure to leave the blocking state and at least equal to the holding current to reach the latched state.

## 3.3 Triggering Modes

Various excitations can produce lateral current flow and trigger latchup if sufficiently large. We shall briefly describe their origin and the corresponding turn-on sequence leading to latch-up. For concreteness the following descriptions pertain explicitly to N-well CMOS unless otherwise noted, but the triggering modes themselves are found in P-well and twin well also.

## 3.3.1 Output Node Overshoot/Undershoot

Holes are injected into the N-well when voltage on an output node connected to a $P^+$ diffusion exceeds the fixed voltage on the N-well, a condition caused, for example, by a signal reflection at a mismatched interface. Holes in the N-well are collected by the reverse biased N-well/substrate junction and enter the substrate. This majority carrier current then produces an ohmic drop in the substrate. If large enough, such a drop can in turn cause a grounded $N^+$ diffusion in the substrate to forward bias.

Reflections at a mismatched interface can also cause an output node to fall below ground. The $N^+$ diffusion tied to the output then injects electrons into the substrate. These electrons diffuse to the reverse-biased N-well where they are collected. The resulting majority carrier current in the N-well produces an ohmic drop in the N-well, which if sufficiently large can cause a $P^+$ diffusion in the N-well to forward bias. This injection can be eliminated by using a substrate bias greater than any expected noise pulse.

The possibility of voltage surges at the input and/or output pins causing latchup has been recognized for some time. (See [Kyomasu-78] for example.) Iizuka and Moll distinguish between two types of latchup associated with overshoot at the output node of a CMOS inverter [Iizuka-81]. When the latchup path is from an output node to ground (for overshoot) or to the power supply voltage $V_{dd}$ (for undershoot), they refer to latchup as "output" latchup (OLU). When the latchup path is from the power supply to ground, they refer to latchup as "main" latchup (MLU). They derive criteria for both types and note that, depending on parasitic transistor parameters, either main latchup or output latchup occurs at the lower trigger current. Output latchup may or may not subsequently trigger main latchup. Output latchup is temporary when the output node supplies current only during a transient. Main latchup can be a more serious problem since it is always sustained and is likely to cause fusing of an aluminum line.

### 3.3.2 Input Node Overshoot/Undershoot

If an input line is connected to a $P^+$ diffusion, as in the double diode protection circuit for example, an overshoot exceeding the power supply voltage causes the injection of holes into the N-well, with the same possible consequences as in 3.3.1.

Other types of protection circuits use various forms of a gated diode ($N^+$ to substrate for N-well CMOS), so that avalanching from overshoots in excess of the reverse breakdown causes hole current to flow in the substrate. If the resulting ohmic drop in the substrate is large enough, a neighboring $N^+$ diffusion can be forward biased, causing electron injection into the substrate.

Since an $N^+$ to substrate diode is used for both types of protection circuits described above, undershoot at the input node has the same result as undershoot at an output node. Again the consequences are discussed in 3.3.1.

### 3.3.3 Avalanching N-well Junction

When both parasitic bipolars are off, the entire power supply voltage is dropped across the reverse-biased N-well/substrate junction. If this voltage is raised sufficiently to cause avalanche, the resulting current flows through both bypass resistors. For a non-retrograded well this breakdown usually occurs at the surface, and its current path can be modeled by a voltage-dependent current source from collector to base of the lateral NPN, as indicated by the current source $I_{avs}$ in Figure 3-1. For retrograded wells in epitaxial CMOS, a second possible avalanche path from the well to the substrate is indicated by the current source $I_{avb}$. When the retrograde well is accompanied by a thin epitaxial layer, the well and out-diffusing substrate form a PN junction with relatively high doping on both sides. Usually the device with the larger bypass resistor turns on first (see section 4.4.4). Because the current-voltage characteristic for the avalanching junction can be very steep, the avalanche current could turn on both transistors nearly simultaneously.

Because of its close relationship with avalanche switching employed in early semiconductor controlled-rectifiers (SCR's), avalanche triggering was utilized early for latchup characterization [Gregory-73] and is still widely used. As long as

## Problem Description

internal and external diffusions are placed far enough from the N-well to preclude punchthrough and short channel field device turn-on, latchup resulting from raising the power supply voltage is caused by avalanche triggering.

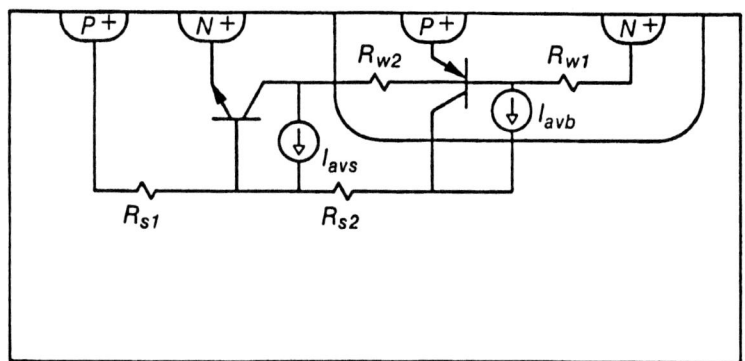

Figure 3-1.  Current path across collector/base junction. $I_{avs}$ models surface avalanche; $I_{avb}$ models bulk avalanche.

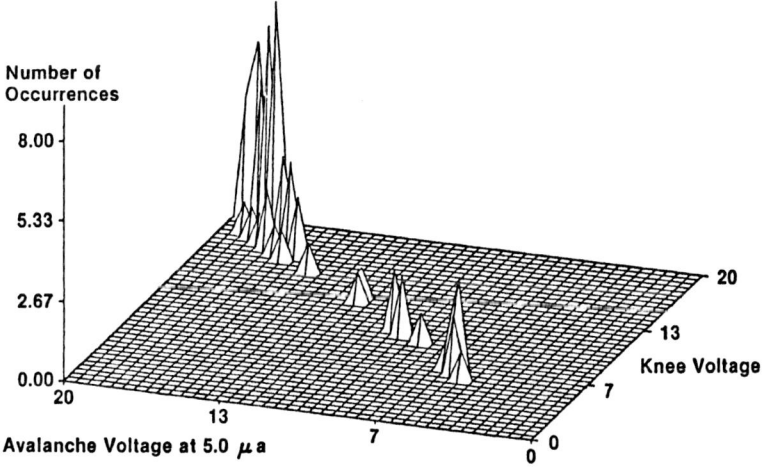

Figure 3-2.  N-well avalanche triggering of latchup.  Knee (switching) voltage is from latchup measurement; avalanche voltage, a separate measurement. ($R_{ss} = R_{wx} = 100$ kΩ; $V_{ss} = 0V$.) From [Craig-84]. © 1984 IEEE. Reprinted with permission.

Figure 3-2 shows test results presented by Craig that illustrate the correlation between N-well avalanche voltage and switching voltage for samples in which the $N^+$ diffusion forming the N-well contact abuts the N-channel field doping [Craig-84]. Substrate bias was zero volts, and 100 kΩ resistors were externally connected to both N-well and substrate to heighten the sensitivity of these measurements.

### 3.3.4 Punchthrough from N-well to External N-Type Diffusion

If an external $N^+$ diffusion is near the N-well, raising the power supply voltage could initiate punchthrough at a voltage lower than that required for avalanching the N-well/substrate junction. (See, for example, [Wollesen-83] or [Craig-84].) As N-well voltage is increased, the junction depletion region can spread into a closely spaced $N^+$ diffusion. The result is the punchthrough current path indicated by $I_{pts}$ in Figure 3-3. Punchthrough voltage is lowest when there is no reverse bias between the external $N^+$ diffusion and substrate. The resulting punchthrough current flowing through the N-well, if large enough, can then turn on the VPNP.

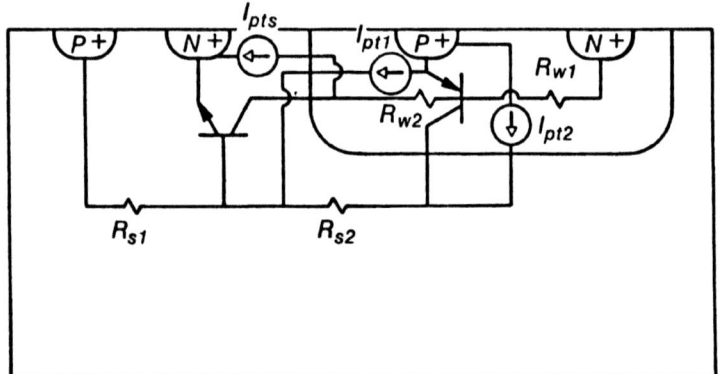

Figure 3-3. Current path from collector to emitter. $I_{pts}$ models punchthrough to an external diffusion. Internal punchthrough is modeled by $I_{pt1}$ for lateral punchthrough and by $I_{pt2}$ for vertical.

Figure 3-4 shows test results presented by Craig that illustrate the correlation between punchthrough and switching voltage for samples having inadequate field doping [Craig-84]. Substrate bias was zero volts, and 20 kΩ resistors were externally connected to

## Problem Description

both N-well and substrate to heighten the sensitivity of these measurements.

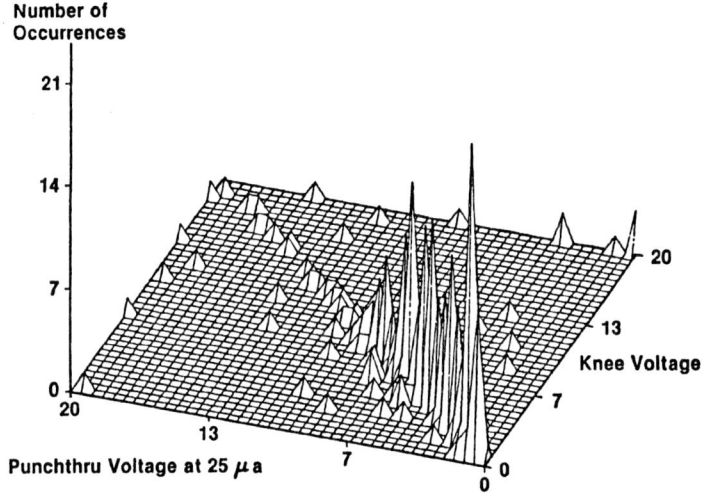

**Figure 3-4.** Punchthrough triggering of latchup. Knee (switching) voltage is from latchup measurement; punchthrough voltage, a separate 3-terminal measurement. ($R_{sx} = R_{wx} = 20$ k$\Omega$; $V_{ss} = 0V$.) From [Craig-84]. © 1984 IEEE. Reprinted with permission.

### 3.3.5 Punchthrough from Substrate to Internal P-Type Diffusion

An alternative punchthrough path to that above, that can occur with increasing power supply voltage (or with decreasing distance from an internal P+ diffusion to N-well edge), is from the substrate to an internal P+ diffusion that is being held at the well voltage. The result is the punchthrough current path indicated by $I_{p t1}$ in Figure 3-3. For a shallow, non-retrograded well the punchthrough path indicated by $I_{p t2}$ is also possible [Lewis-84]. Such paths are all the more likely if there is no closely spaced external N+ diffusion because the N-well/substrate depletion region spreads back into the relatively lightly doped N-well in addition to spreading downward into the substrate and laterally into the epitaxial layer. The resulting punchthrough current, if large enough, can turn on the LNPN.

This vertical punchthrough has been cited as a major design consideration for minimum well dose [Schwabe-83], and it has in fact been observed to cause latchup during overvoltage stress in shallow P-well CMOS when the epitaxial layer thickness is 4 μm or less [Lewis-84]. Leventhal also reports such a case of punchthrough at 70 volts during an attempt to cause triggering by N-well/substrate breakdown [Leventhal-84]. For the sample being tested, punchthrough to an internal $P^+$ diffusion (corroborated by a separate measurement) occurred prior to avalanching in the N-well junction.

### 3.3.6 Parasitic Field Devices

A parasitic FET is formed in the field region between the N-well, which acts as a drain, and a closely spaced external $N^+$ diffusion. This parasitic N-channel FET can be either gated or ungated, with oxide charge causing channel current in the latter case. The effect of a parasitic N-channel device when it turns on is similar to punchthrough, and its effect can be represented by a current source in parallel with $I_{pts}$ in Figure 3-3. Parasitic channel current through the N-well then tends to turn on the vertical PNP. Technically, the lateral NPN is already on if FET current is flowing since the $N^+$/substrate junction is already forward biased at the surface by the gate of the parasitic FET.

Likewise, a parasitic P-channel FET is formed between an internal $P^+$ diffusion and the substrate, which acts as a drain. When the parasitic channel turns on, channel current through the substrate tends to turn on the lateral NPN. Its effect can be represented by a current source in parallel with $I_{pt1}$ in Figure 3-3. Technically, the vertical PNP is already on since the $P^+$/N-well junction is already forward biased at the surface by the gate of the parasitic FET.

This triggering mode has been extensively studied by Takacs, Werner, Harter, and Schwabe using a PNPN structure with a gate electrode over the field oxide and between the internal and external parasitic emitters [Takacs-82,84]. Figure 3-5 shows illustrative results for P-well CMOS. At low gate voltage, $V_{crit}$ (their designation for switching voltage) rises with a slope of slightly less than a volt per volt. In this region the effective gate-to-source voltage -$(V_{crit} - V_{gs})$ is more negative than the P-channel field threshold, and latchup is initiated by hole current in the P-channel field device. When a bias is applied to the P-well,

## Problem Description

a larger hole current is required to turn on the VNPN, and $V_{crit}$ must increase to provide the necessary drive. The sharp drop in critical voltage at high gate voltage results from the relatively sharp turn-on of the N-channel field device. Its drain current causes an ohmic drop in the substrate, turning on the LPNP. Applying a bias to the P-well provides a source/substrate bias and raises the N-channel field threshold. In the authors' opinion, field oxide short channel effects pose the most severe limitation to latchup immunity at small $N^+/P^+$ spacings.

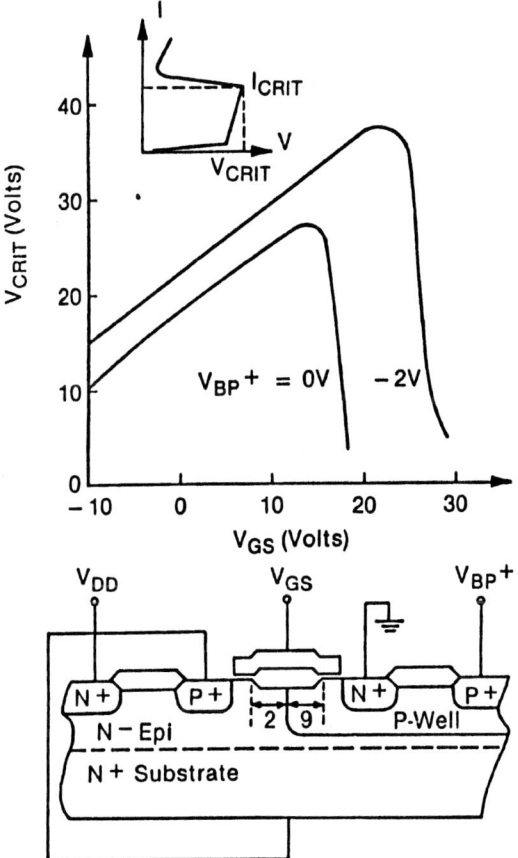

Figure 3-5. Field FET triggering of latchup. Field electrode voltage influences switching voltage. See text for explanation. From [Takacs-84]. © 1984 IEEE. Reprinted with permission.

The dynamics of this triggering mode have been studied by Odanaka, Wakabayashi, and Ohzone using two-dimensional numerical simulation [Odanaka-85]. Supply overvoltage was used to trigger latchup with the poly silicon P-channel field electrode held at ground. Hole and electron currents at the well/substrate junction were simulated vs. time for $V_{dd}$ = 7.9 V (no latchup) and for $V_{dd}$ = 8.2 V (latchup occurs). Hole current begins first, indicating that in their sample channel formation caused by drain-induced barrier lowering in the field P-channel FET occurs at lower $V_{dd}$ than does punchthrough to an external $N^+$ diffusion. Hole current quickly rises to greater than 100 mA/cm, continues a slow rise until 3.5 nS after $V_{dd}$ is applied, then rises sharply as latchup ensues. Hole current through the substrate creates an electric field that causes field-aided base transport in the LNPN. Electron current begins to rise after 0.4 nS, slowly at first, then sharply after 3.5 nS. Both hole and electron currents flow near the surface, and their distributions broaden with time and current amplitude.

### 3.3.7 Photocurrent

Satellite applications guided early characterization of CMOS circuits, and experimental transient measurements were performed on a variety of commercially available CMOS MSI chips to assess their latchup hardness [Gregory-73]. Various forms of radiation can produce hole/electron pairs throughout the silicon substrate. Photo-induced holes produce majority carriers in the substrate, and the resulting current tends to turn on the lateral NPN. Likewise, photo-induced electrons produce majority carrier current in the N-well, tending to turn on the vertical PNP.

Photo current is modeled by means of a current source across the N-well/substrate junction. In the blocking state it flows through both base/emitter bypass resistors of a PNPN structure [Estreich-78]. Which transistor turns on first depends on the bypass resistor values. A model for the strength and time dependence of the current generator for a pulse of ionizing radiation absorbed uniformly throughout the silicon is given in [Estreich-80].

Problem Description                                                33

### 3.3.8  Avalanching Source/Drain Junctions

An avalanching $N^+$ drain produces substrate hole current which can forward bias a nearby source. If the resulting ohmic drop in the substrate is large enough, this can lead to snapback, even in NMOS technologies [Kennedy-73]. If the source is turned on hard enough, a sufficient number of injected electrons might be collected by the N-well (some fraction of them are collected at the N-channel FET drain) to turn on the vertical PNP.

An avalanching P-channel drain produces an N-well current that tends to forward bias the $P^+$ source. This can also lead to snapback, although the problem for P-channel devices is not as serious as for N-channel, because of an inherently more gradual doping gradient for the $P^+$ diffusions (resulting in a lower peak electric field) and because of a lower ionization rate for holes. When the source is turned on, injected holes are collected by the N-well junction and flow through the substrate to be collected at a substrate contact. A sufficiently large substrate current could then turn on the lateral NPN.

### 3.3.9  Displacement Current

A ramped power supply or noise on the power supply line causes a displacement current through the RC circuit formed by the two bypass resistors and the N-well/substrate junction capacitance. This displacement current has been modeled using a current generator across the well/substrate junction [Estreich-78], and it tends to first turn on either the lateral NPN or the vertical PNP, depending on which has the larger bypass resistor. Actual circuit layouts exist in which one, or both, of the bypass resistors is partially, or wholly, shunted by a parasitic coupling capacitor, in which case the bypass impedance is lowered. Troutman and Zappe have shown the general condition governing these cases to be that displacement current excitation must raise PNPN terminal current to the switching current in order to cause latchup [Troutman-83a]. They also investigated the dynamics of displacement-current triggered latchup using piecewise linear transistor models to obtain a closed form analytic solution.

## 3.4 Triggering Taxonomy

The above triggering modes divide into three types. This taxonomy facilitates application of the latchup criterion developed in Chapter 4, systematizes the characterization techniques discussed in Chapter 5, and focuses the latchup prevention strategies presented in Chapter 6.

### 3.4.1 Type 1: External Initiation of First Bipolar.

In Type 1 triggering some excitation has turned on the first transistor. This excitation is usually an overshoot or undershoot at an input and/or an output node. Latchup follows if the second transistor is turned on by the first and if total current into the parasitic PNPN reaches the switching current calculated for the particular PNPN triode structure that results when the bypass resistor on the first transistor is removed. Latchup is sustained if the total current has also reached the switching current for the full tetrode structure. If the total current lies between the switching currents for the two configurations, latchup is temporary, and the PNPN device returns to the blocking state once the external source is removed. As will be pointed out in Chapter 4, if one bypass resistor exceeds the other by more than a factor of two, the switching current for the tetrode configuration can be approximated rather well by using the triode configuration obtained by removing the larger bypass resistor. Switching current for the two triode configurations is essentially identical (and latchup is always sustained) if the excitation has turned on the parasitic bipolar with the larger bypass resistor because the two triode approximations are then identical.

### 3.4.2 Type 2: Normal Bypass Current Initiation of Both Bipolars

In type 2 triggering the excitation causes current to flow through both bypass resistors. The excitation in this case is usually avalanche current, photocurrent, or displacement current through the well/substrate junction. Usually the transistor with the larger bypass resistor turns on first (see section 4.4.4). For latchup to occur, the second transistor must also turn on, and PNPN current must reach the switching current as calculated for the full tetrode configuration. (Again, the appropriate triode configuration yields an excellent approximation if the bypass resistors differ by more than 2X.) Latchup is now always sustained.

### 3.4.3  Type 3: Degraded Bypass Current Initiation of Both Bipolars

In Type 3 triggering the excitation has already degraded the blocking state by creating a lower impedance path. This path is between the power supply and ground or between power supply and substrate supply (or substrate generator) in the case of punchthrough or a field FET device. In the case of source/drain avalanche, it is between the power supply and a signal line or between a signal line and and the substrate supply (or substrate generator). This degradation is itself undesirable and should be avoided in the process design. Now the first "transistor" is not exhibiting transistor action, although under the influence of an external source its structure is supplying current that, if large enough, can turn on the second transistor. The source of this current is punchthrough, a field FET, or drain junction avalanche. At least temporary latchup occurs once the PNPN current has reached the switching current calculated for the triode configuration formed by removing the excited transistor's bypass resistor. Whether latchup is sustained depends on the same set of conditions as for Type 1 triggering.

### 3.4.4  Latchup Sequence and Classification Summary

Figure 3-6 summarizes turn-on sequences leading to latchup for the various excitations. To make the taxonomy more specific, this table has been compiled for N-well CMOS. The first entry is an example of type 1 triggering. An external signal directly excites the parasitic vertical PNP by applying a forward bias on the base/emitter junction. The resulting VPNP collector current flowing through $R_s$, if large enough, excites the parasitic lateral NPN bipolar transistor.

In type 2 triggering the latchup sequence depends on which bypass resistor is larger. Since the same current flows through both bypass resistors, the transistor having the larger bypass resistor turns on first. The same sequences are observed in type 3 triggering, but now the same current does not flow through both bypass resistors.

| Type | Triggering Mode | Latchup Sequence |
|---|---|---|
| 1A | I/O Node Overshoot | V → VPNP → $R_s$ → LNPN |
| 1B | I/O Node Undershoot | V → LNPN → $R_w$ → VPNP |
| 2A | Avalanching N-well | If $R_w > R_s$, then<br>I → $R_w$ → VPNP → $R_s$ → LNPN |
| 2B | Photocurrent | |
| 2C | N-well displacement current | If $R_s > R_w$, then<br>I → $R_s$ → LNPN → $R_w$ → VPNP |
| 3A | External Punchthrough | I → $R_w$ → VPNP → $R_s$ → LNPN |
| 3B | Internal Punchthrough | I → $R_s$ → LNPN → $R_w$ → VPNP |
| 3C | N-channel field FET | I → $R_w$ → VPNP → $R_s$ → LNPN |
| 3D | P-channel field FET | I → $R_s$ → LNPN → $R_w$ → VPNP |
| 3E | Avalanching $P^+$ diffusion | I → $R_w$ → VPNP → $R_s$ → LNPN |
| 3F | Avalanching $N^+$ diffusion | I → $R_s$ → LNPN → $R_w$ → VPNP |

Figure 3-6. Triggering taxonomy and latchup sequence.

# Chapter 4

## LATCHUP MODELS AND ANALYSES

The previous chapter showed us how the latchup sequence can be used to categorize triggering modes and reduce the apparent complexity of our investigation. Looking back at the territory already covered in our exploration supplies us with some additional perspective. Mindful of previous explorers, we also want to utilize their trail blazes wherever possible.

Chapter 2 identified the origin of the PNPN structure inherent to bulk CMOS and introduced the lumped element model. As simple as that model appears, the literature contains neither a general analysis of its behavior nor a comprehensive latchup criterion derived from it. This chapter provides both. It also provides a rigorous and general technique for predicting blocking state behavior, including the location of the switching boundary, and for guiding experimental observations and characterization procedures. The key to understanding this new latchup criterion, as well as most latchup experiments, is knowing whether a parasitic bipolar operates in the normal or bypassed mode when it is turned on.

Although the lumped element model has provided a useful level of abstraction for understanding latchup and for assessing latchup hardness of integrated PNPN structures, establishing values for the lumped elements is often complicated by two (or even three) dimensional effects. This chapter solves this problem for distributed resistances and demonstrates how the transmission line model can easily relate the equivalent resistance to doping profiles and layout dimensions.

Recent work has initiated explorations of dynamic latchup behavior and has set lower bounds on the triggering duration required for switching. Two-dimensional simulations have graphically illustrated the changing potential and charge carrier distributions as the parasitic PNPN structure switches from the blocking to the latched state. These simulations have also revealed unexpected current flow patterns when the two parasitic bipolars are interacting.

## 4.1 Antecedents of Latchup Models

Before deriving a new, comprehensive latchup criterion and closely examining latchup behavior, it will prove instructive to trace the evolution of the lumped element model for the parasitic PNPN structure. We begin with its antecedents from the early years of semiconductor devices, then continue with a historical review that examines its equational legacy developed from years of service describing CMOS latchup.

### 4.1.1 Hook Collector Transistor

Initial modeling of a four layer PNPN structure originated with an attempt to understand early type-A point contact transistors. In those transistors the emitter and collector were fabricated by alloying wire to the surface of N-type germanium to create local p-regions. One might then have expected a "lateral" PNP transistor to result whose common-base current gain, although dependent on the fabrication conditions, was certainly less than unity. When alloying was performed using electrical pulses, however, it was found that in many cases the current gain greatly exceeded unity. Changes in emitter current could produce changes in collector current one to two orders of magnitude larger. The explanation offered was that one point contact heated more than the other, causing thermal conversion to take place [Shockley-50, 51]. Instead of a PN diode, this specially formed collector contact was a three layer NPN structure with the p-region floating, as shown in Figure 4-1.

In the resulting PNPN transistor the p-type emitter was positively biased with respect to the grounded n-type base, and the other n-region was biased negatively so that the middle PN junction was reverse-biased. Holes from the p-type emitter that

## Latchup Models and Analyses

reached the reverse-biased collector were trapped by the "hook" in the potential energy curve caused by the floating p-region. This trapped charge forward biased the emitter/base junction of the NPN structure, causing electrons to be injected into the floating p-region. After diffusing through the p-region, these electrons also were collected by the reverse-biased junction. In transistor parameter terminology, some fraction $\alpha_{fp}$ of the PNP emitter current consisted of holes that reached the floating collector where, because this floating p-region was also the base of an NPN, they caused an electron current of $1/(1 - \alpha_{fn})$ to flow, $\alpha_{fn}$ being the current gain of the NPN in a common base configuration.[1] Thus, the overall current gain of the PNPN transistor was $\alpha_{fp}/(1 - \alpha_{fn})$, which clearly can be much larger than unity.

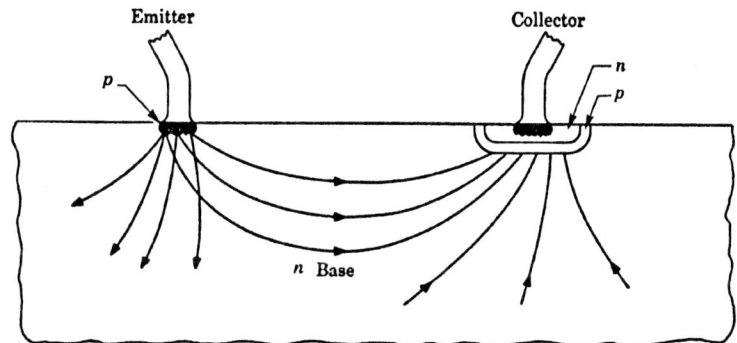

Figure 4-1. Point-contact transistor. The collector region includes a floating p-type base. From [Lindmayer-65]. © 1965 D. Van Nostrand Co. Reprinted with permission.

Creating the first equivalent circuit for the PNPN transistor, Ebers analyzed the dependence of the overall alpha on emitter current [Ebers-52]. In addition to the normal 3-terminal hook collector configuration, in which the p-region forming the PNP collector (and NPN base) is left floating, he considered a 4-terminal configuration in which this p-region is connected

---

[1] Throughout this book transistor alphas have their usual meaning. The forward alpha $\alpha_f$ is the product of emitter-to-base injection efficiency times the base transport factor from emitter to collector, and the reverse alpha $\alpha_r$ is collector-to-base injection efficiency times the base transport factor from collector to emitter.

externally to the NPN emitter via either a small resistor or a diode with its polarity opposite to that of the NPN emitter/base junction. He demonstrated both theoretically and experimentally (using a combination of an NPN and a PNP transistor) that the latter configuration could be used to reduce the gain and make it more uniform as emitter current varied.

Using an emitter/base bypass resistor to reduce transistor gain is a key concept in hardening bulk CMOS against latchup, and we will return to this subject in later sections and quantify the efficacy of bypassing one or both of the base/emitter junctions.

### 4.1.2 Semiconductor Controlled Rectifier (SCR)

The next modeling effort for the PNPN structure occurred during the development of the semiconductor controlled rectifier (SCR), today more commonly referred to as a thyristor. As an SCR the PNPN structure functions as a switching device. Its two stable states (ON and OFF), each with low power consumption, make it uniquely suited to applications requiring latching action and power handling capability, such as AC to DC power conversion, electric motor drives, and the control of electrical servomechanisms.

In the OFF state, the middle PN junction is reverse-biased and blocks any current flow (except for leakage current). To trigger the SCR into the ON (or conducting) state, the blocking junction must be discharged. In general terms this is accomplished by increasing the concentration of majority carriers adjacent to the blocking junction. When majority carriers are supplied to the base of a transistor in excess of emitter efficiency and minority carrier recombination requirements, a buildup of carriers occurs in the base region, causing additional forward biasing of the emitter/base junction with an attendant rise in emitter current. We saw one manifestation of this behavior in the hook collector transistor, in which the collector current of one transistor flooded the base region of the second.

In early PNPN devices operated as controlled rectifiers, both bulk and depletion lifetimes tended to be extremely short. As a result, current gain in the composite transistors was poor, and most early analyses realistically assumed $(\alpha_{fn} + \alpha_{fp}) < 1$ at low current levels. In Moll, Tannenbaum, Goldey, and Holoynak for example, the I/V characteristics are derived for a 2-terminal (both internal layers are floating) or "diode" configuration of the PNPN structure, which is sometimes referred to as a Shockley diode [Moll-56]. Their result is

$$I = \frac{I_{s2}}{[1 - (\alpha_{fn}M_n + \alpha_{fp}M_p)]}, \qquad (4-1)$$

where $I_{s2}$ is the current that would flow if the middle junction were isolated and reverse biased; $\alpha_{fn}$ and $\alpha_{fp}$ are the forward alphas for the NPN and PNP transistors; $M_n$ and $M_p$ are electron and hole multiplication factors, and each is a function of the reverse bias $V_2$ on the middle junction. Applying positive voltage to the anode initially produced a high impedance state because the middle junction was reverse biased and blocked any current flow. Raising the anode voltage eventually switched the SCR diode to the low impedance ON (or latched) state, and one observed a breakover in the I/V characteristics. This breakover voltage occurs when the product of junction avalanche multiplication and low current alphas is equal to unity. (Although not explicitly stated by the authors, small-signal values should be used for this condition, as will become clear in subsequent paragraphs.) In this case the majority carrier concentration adjacent to the blocking junction is increased by avalanche multiplication.

Their paper also includes a derivation of the Shockley diode current/voltage characteristics for the low impedance state. For ON state currents greatly exceeding the saturation currents of all three PN junctions, this characteristic is well approximated by the expression

$$I = A[\frac{I_{s1}I_{s3}}{I_{s2}}]e^{V/V_t}, \qquad (4-2)$$

where $V_t = KT/q$ is the thermal voltage and $I_{s1}$, $I_{s2}$, and $I_{s3}$ are the saturation currents for each of the junctions when the other two are short circuited (junction 2 being the well/substrate junction and junctions 1 and 3 the two emitter/base junctions). The coefficient A is given by

$$A = \frac{(\alpha_{fn} + \alpha_{fp} - 1)(1 - \alpha_{fn}\alpha_{rn} - \alpha_{fp}\alpha_{rp})}{[(1 - \alpha_{fn}\alpha_{rn}) - \alpha_{rp}(1 - \alpha_{fn})][(1 - \alpha_{fp}\alpha_{rp}) - \alpha_{rn}(1 - \alpha_{fp})]},$$

(4 – 3)

where $\alpha_{rn}$ and $\alpha_{rp}$ are the reverse alphas for the NPN and PNP.[2] Since the individual transistors are driven into saturation, the characteristic for the low impedance state is functionally dependent on their reverse alphas, as well as their forward alphas and saturation currents.

On the other hand, if $(\alpha_{fn} + \alpha_{fp}) \geq 1$ at all current levels, there is no forward blocking state for the diode configuration. Only one state (the low impedance state) exists, and the transistors operate simultaneously in both forward and reverse mode as soon as voltage is applied. Then the general characteristics presented in [Moll-56], rather than the approximation shown above, must be used for describing Shockley diode behavior over the entire current range.

Mackintosh extended PNPN device analysis to the triode configuration, in which both emitters and one base are electrically connected [Mackintosh-58]. The external lead connected to the base is referred to as the gate, but gate current equals base current only when collector current for the other transistor is zero. Again it is assumed that $(\alpha_{fn} + \alpha_{fp}) < 1$ at low current, and gate current is used to increase one transistor's gain so that breakover can be achieved without the need for avalanche multiplication at high anode voltage. As can be seen in Figure 4-2, using gate current to control rectification in the PNPN device reduces both breakover voltage and turn-off current, the current at which bias on the

---

[2] Because there is only one emitting junction (J2) for the reverse mode, the reverse alphas satisfy the constraint $(\alpha_{rn} + \alpha_{rp}) \leq 1$. The equality holds if the base transport factors for holes and electrons are both unity.

middle junction is zero. Positive gate current is used to switch the PNPN device from the high to the low impedance state, and negative gate current is used to switch back from low to high. Mackintosh presents a general derivation for the forward characteristics of the PNPN triode at constant gate current, including a breakover condition and an equation for turn-off current. As he acknowledges, his equations are not valid for the ON region. However, because he makes the simplifying assumption that the triode's electrical behavior can be explained by reference to the voltage across the center junction alone, he wrongfully concludes that the turn-off point actually separates the ON region from the negative differential resistance region. As shown in section 4.6.1, this separation actually occurs at the holding point, where the collector/base junction is heavily forward biased, and holding current is much larger than turn-off current.

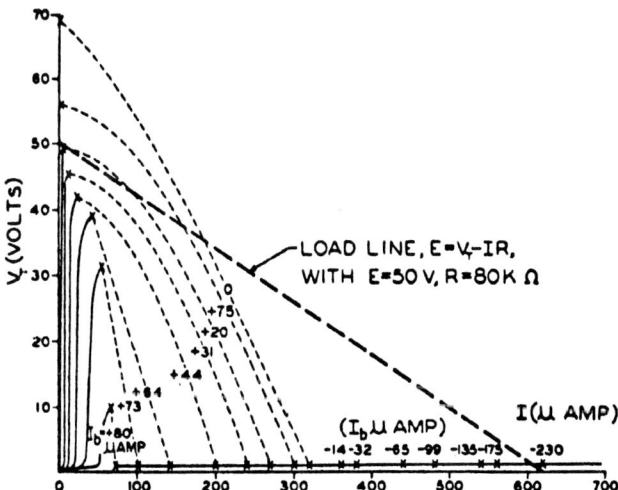

Figure 4-2. Experimental V/I characteristics for SCR triode. Positive base current enhances switching into low impedance state. From [Mackintosh-58]. © 1958 IRE (now IEEE). Reprinted with permission.

In later work (reviewed in [Gentry-64]), other triggering techniques were used to raise the transistors' emitter currents, and consequently their current gain, to the point at which the SCR switches from the high to the low impedance state. These triggering mechanisms included thermal turn-on (leakage current rises exponentially with temperature), light or radiation triggering, and voltage (or dV/dt) triggering. Early in the development of the SCR it was realized that switching was faster, required less power, and occurred more predictably when a gate was added to the four layer structure so that triggering need not rely on avalanching the blocking junction. Various methods for triggering the parasitic PNPN in modern CMOS are covered in Chapter 3. Here we continue our review of SCR switching criteria.

The book by Gentry, Gutzwiller, Holoynak, and Von Zastrow extensively discusses the triode configuration, which can be switched more quickly and more efficiently than the diode configuration because lower voltages are involved [Gentry-64]. When gate current $I_g$ is forced into the base of the NPN, the anode current for the case $(\alpha_{fn} + \alpha_{fp}) < 1$ can be expressed as

$$I_a = \frac{\alpha_{fn} I_g + I_{s2}}{1 - (\alpha_{fn} + \alpha_{fp})} . \qquad (4-4)$$

Increasing $I_g$ causes the alpha sum to increase to some critical value that causes switching into the low impedance state. This occurs not when $\alpha_{fn} + \alpha_{fp} = 1$, as might be suggested by the above equation, but at a somewhat smaller sum determined by incremental changes of gate current. As the authors clearly demonstrate, the change in a triode's anode current caused by a change in gate current is given by

$$\frac{dI_a}{dI_g} = \frac{\alpha_{fns}}{1 - (\alpha_{fns} + \alpha_{fps})} , \qquad (4-5)$$

where the small-signal alphas are defined by $\alpha_{fs} = dI_c/dI_e$. Thus, gated switching in the triode configuration occurs when $(\alpha_{fns} + \alpha_{fps}) \geq 1$.

They also point out that for the case $(\alpha_{fn} + \alpha_{fp}) > 1$, gate current can be used to provide a blocking state, so that two states are possible in the triode configuration throughout the full range of values for the alpha sum. To maintain the blocking condition, the gate current is held below some critical value (or even made negative). As such it diverts current either around the cathode (usual case for the PNPN triode), around the anode, or around both anode and cathode (the tetrode configuration). This is the key idea in controlling latchup since for modern CMOS technology it is generally true that the sum of forward alphas for the parasitic transistors exceeds unity, in which case a negative gate current is required for blocking. This case is so generic to latchup situations that we shall refer to the negative gate current as bypass current $I_{by}$ (so that bypass current is defined to be positive when flowing in the same direction as the emitter current of the bypassed transistor). When we analyze the parasitic PNPN in a later section, we shall use their concept of effective injection efficiency, defined for the blocking state ($I_{by} \geq 0$) as

$$\gamma^* = \frac{I_e}{I_e + I_{by}}, \qquad (4-6)$$

where $I_e$ is emitter current and $I_{by}$ is the corresponding bypass current. The sum $I_e + I_{by}$ equals the total current I through the PNPN device when it is in the blocking state.

Gibbons reviews switching criteria applicable to diode, triode, and tetrode configurations of the PNPN structure [Gibbons-64]. He notes that there are two points on the I/V characteristic defined by the condition dV/dI = 0. The first, occurring at the lower current, is the switching point while the second defines the holding point. The switching current is the largest current that can statically hold the PNPN in the high impedance state, and the holding current is the smallest current that can statically hold the PNPN in the low impedance state. The region between these two points exhibits a negative differential resistance.

Gibbons notes that deriving the general expression corresponding to dV/dI = 0 is difficult and instead offers approximate analytical expressions describing the two points. He first derives an equation for the voltage $V_2$ across the blocking junction in terms of transistor parameters and of hole and electron multiplication factors, as well as the current I. As an approximation to the switching point he uses the condition $dV_2/dI = 0$ to demonstrate that if $(\alpha_{fn} + \alpha_{fp}) < 1$ at low current, switching occurs when

$$M_n \alpha_{fns} + M_p \alpha_{fps} = 1. \qquad (4-7)$$

The derivatives $dV_1/dI$ and $dV_3/dI$ are both positive when this condition holds. This equation[3] implicitly determines the coordinates $(V_2, I)$ at which the center junction has zero dynamic impedance. At larger currents $dV_2/dI$ is negative, and $V_2$ decreases. As an approximation to the holding point he uses the condition $V_2 = 0$, the point at which the center junction is about to become forward biased (sometimes called the turn-off point). This leads to the equation for total current into the PNPN device at the turn-off point, namely,

$$I_{to} = \frac{\alpha_{fn} I_{g1} + \alpha_{fp} I_{g2}}{1 - (\alpha_{fn} + \alpha_{fp})}, \qquad (4-8)$$

where $I_{g1}$ and $I_{g2}$ are gate currents defined as positive when current flows from the PNP base and to the NPN base, respectively.[4] He is careful to point out that $V_2 = 0$ does not really define the holding point, since the middle junction is actually heavily forward biased at the holding point. As a quantitative measure of this difference, he solves the special case of idealized "identical" NPN and PNP transistors and shows the holding current to be 2.5 times the turn-off current.

It is tempting to believe the above equations describing SCR switching, such as (4-7) and (4-8), carry over to CMOS parasitic PNPN structures. However, one must be careful. Because of high quality wafers and careful fabrication steps in modern CMOS

---

[3] Macintosh's breakover condition for the triode is easily shown to have the same form.

[4] Setting $I_{g2} = 0$ yields Macintosh's equation for triode turn-off current.

technologies, the parasitic bipolar transistors often have excellent low current characteristics, with $(\alpha_{fn} + \alpha_{fp}) > 1$ for base currents down to the level of emitter/base saturation current. We have already seen that bypass resistors are required to provide a blocking state for such PNPN structures, and modifications to the switching equations are also necessary. The derivation of these equations and discussion of their application is rather lengthy and will be presented in later sections. We now turn our attention to the lumped element model for the PNPN structure and review its application to CMOS.

## 4.2 Evolution of the PNPN Lumped Element Model

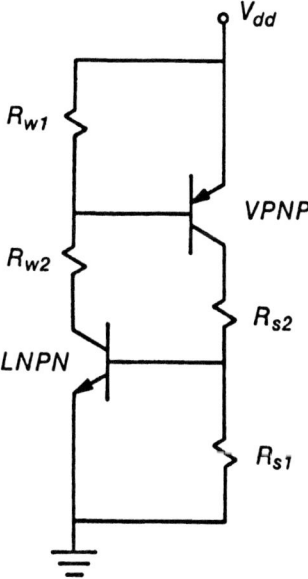

Figure 4-3. Lumped element model for 4-terminal PNPN structure. This is the most common version.

Although the lumped element model does not always give a complete description of latchup, it has proved immensely useful for discussing latchup fundamentals and guiding CMOS process development. In its various forms it has also proved remarkably versatile in explaining, at least qualitatively, the diverse experimental characteristics observed for latchup. Over time the model, and the means for extracting its parameter values, have undergone several extensions and refinements. We now review the major items in the evolution of the lumped equivalent model by discussing its individual components and how they affect latchup behavior.

Because of its importance in subsequent discussions, the schematic for the PNPN structure is shown redrawn for reference in Figure 4-3. Here the lateral NPN emitter and bypass resistor are shown explicitly connected to ground, which is the usual case for digital circuits. For clarity, the present work adheres to this convention unless otherwise noted. In analog circuits a separate power supply $-V_{ss}$ is frequently used to provide for equal positive and negative voltage swings around ground.

### 4.2.1 Resistive Elements

Usually when parasitic bipolars form PNPN structures, the internal spreading resistance of the well and substrate shunts the base/emitter junctions. Gregory and Shafer have pointed out that the presence of these shunt resistors in normal circuits is crucial for understanding latchup [Gregory-73]. Lateral current flow through these resistors is generally the mechanism causing the parasitic bipolars to turn on. Although they noted less latchup sensitivity to ionizing radiation when these resistors were in place, they provided no quantitative measure of their effect. In addition, the gain condition they cite as necessary for latchup shows no dependence on these resistors.

All subsequent latchup modeling has included these shunt or bypass resistors unless the well and substrate contacts were left unconnected. Calculating the values for the lumped equivalent resistors has sometimes been a problem, however, and we shall return to this point in section 4.7.

Dressendorfer and Ochoa observed both large holding voltage offsets and saturation of the total current through the PNPN structure on some samples [Dressendorfer-81]. This behavior is not normally present in discrete SCR's or Shockley diodes, and they attributed their observations to distributed resistance effects on their CMOS wafers. In particular, series collector resistance can cause large offsets in holding voltage. Such resistance is common for the parasitic vertical bipolar in non-epi CMOS technologies and for the lateral parasitic bipolar in general because the parasitic emitter in the well must be located some distance from the well's vertical junction.

To make the discussion of the resulting effects more explicit, let us assume collector resistance for the VPNP to be much greater than that for the LNPN, i.e., $R_{cp} >> R_{cn}$, as is the case for a PNPN structure in non-epi CMOS for example. Then, when the circuit latches, the VPNP goes into saturation while the LNPN remains in the active mode because the $IR_{cp}$ drop holds a reverse bias on the collector/base junction of the LNPN. Part of the well/substrate junction is now partly forward-biased, partly reverse-biased. (Such a situation is illustrated in Figure 4-26 on page 110, which simulates an instant during the latchup transient.) As total current increases (with at least one transistor in saturation), current gain of individual transistors declines because of high level injection. Eventually a point is reached at which the collector/base voltage goes to zero, and the device previously in saturation enters the active region. A small increase in current then requires a relatively large increase of collector-to-emitter voltage $V_{ce}$ to maintain the loop gain at unity since the active transistor's gain is relatively insensitive to $V_{ce}$. The same authors have also noted that obtaining the characteristics for the individual parasitic transistors appropriate to the latched condition can be difficult since current flow patterns when the circuit is latched can be very different from those when transistor characteristics are measured using the standard three terminal configuration.

### 4.2.2 Parasitic Bipolar Behavior

Many standard bipolar characterization techniques have been used to obtain parameter values for the parasitic bipolars, but there are several features that warrant special consideration. The lateral electric field accompanying substrate current (for example, the collector current from the VNPN in a p-well technology) can enhance the LPNP current gain [Estreich-78]. This enhancement results from two sources: (1) the lateral field adds a drift component to the LPNP base current and (2) biasing of the lateral emitter/base junction reduces vertical injection by concentrating injection at the edge nearest the P-well collector. Field-aiding is greatest at large ratios of base width to base diffusion length and is most pronounced when the lateral transistor's emitter is located between the well and the contact to the substrate.

On the other hand, in a retrograde well the field retarding effects can substantially degrade the current gain of the vertical transistor. Most of the base recombination occurs in the retarding field region so that in this case the base transport factor, not emitter efficiency, limits VNPN current gain [Estreich-78].

Many authors have attempted parasitic bipolar characterization using 3-terminal measurements to predict or explain latchup results for the 4-terminal PNPN structure, with varying degrees of success. For example, Schroeder, Ochoa, and Dressendorfer combine the $\beta_n(I_{cn})$ and $\beta_p(I_{cp})$ curves using Kirchoff's current laws to predict whether latchup is possible [Schroeder-80]. If these two resulting curves do not overlap, latchup is not possible, while if they do, the overlap defines the range of collector current over which operation in saturation is possible. Here again the difficulty is that current flow patterns for two bipolars interacting can be significantly different from those occuring for normal 3-terminal bipolar measurements.

Characterization of the lateral device is especially complicated because during latchup its base drive originates at the vertical collector region while during bipolar device characterization an external base contact is commonly used. Ochoa and Dressendorfer suggest a 4-terminal measurement of the lateral $\beta$ to account for sourcing of base current at the collector region of the opposite device [Ochoa-81]. Using such a measurement, which they claim yields a larger value for $\beta$ than do conventional measurements, they are able to account for latchup in a circuit that normal bipolar gain measurements predicted to be free from latchup. Leventhal's measurements for non-epi samples using this technique do not show a large difference in $\beta$ values, however [Leventhal-84].

### 4.2.3 Previous Latchup Criteria

The first concise statement of the conditions necessary for latchup was given by Gregory and Shafer [Gregory-73]. Their three conditions were as follows:

1. The transistors' common-base current gains must satisfy $\beta_{fn}\beta_{fp} \geq 1$ (or, equivalently, $\alpha_{fn} + \alpha_{fp} \geq 1$).

2. Both transistors must be turned on.

3. The power supply and bias circuitry must be capable of supplying an amount of current in excess of the holding current.

Strictly speaking, their first condition is the static latchup condition for a PNPN diode configuration. In that it applies to a diode configuration, it is overly restrictive when applied to the more common tetrode (or even triode) configuration, but in that it uses the static criterion it is overly generous. As given above, the second condition is necessary, but not sufficient. Investigations of dynamic latchup effects have shown that both transistors must be held on for some length of time before latchup occurs. As for the third condition, the power supply need source only the switching current (which is somewhat less than the holding current) for unsustained latchup. Once current through the parasitic PNPN device reaches the switching level, it leaves the blocking state, leading to at least momentarily unpredictable circuit behavior.

Kyomasu, Araki, Ohtsuki, and Nakayama suggested that the first condition above must be changed to account for the base/emitter bypass resistors [Kyomasu-78]. They suggested modifying the alphas to read $\alpha_{mod} = \alpha R_s/(R_e + R_s)$, where $R_e$ is the resistance in series with the emitter and $R_s$ is the resistance bypassing the base/emitter junction. However, for zero series emitter resistance this criterion reduces to the condition that the sum of the normal alphas equals unity independent of the values for the bypass resistors, a clearly incorrect result. As shown in section 4.3.3, the correct result has the same form, but $R_e$ is replaced by the total small-signal emitter resistance, which is a function of emitter current.

Ochoa, Dawes, and Estreich have presented a modification to condition 1 that accounts for the shunt resistors when the power supply current is constrained to some maximum value $I_{Dmax}$ [Ochoa-79]. (Also see [Estreich-80].) Their modification is

$$\beta_{fn}\beta_{fp} \geq 1 + \frac{(\beta_{fn} + 1)(\beta_{fn}I_{rw} + I_{rs})}{\beta_{fn}(I_{Dmax} - I_{rw}) - (\beta_{fn} + 1)I_{rs}}. \qquad (4-9)$$

Note that this condition reduces to that given by Gregory and Shafer if either (a) there is no shunt current (independent of the amount of supply current) or (b) supply current is unlimited (independent of the amount of shunting current). In the more general case where supply current is limited and the shunting resistors are present, this condition says the beta product must exceed a quantity greater than unity before latchup can occur.

Iizuka and Moll have presented a figure of merit for investigating latchup hardness of a CMOS inverter triggered by output node overshoot or undershoot [Iizuka-81]. "Output" latchup occurs when the parasitic bipolar whose emitter is connected to the output node forms a PNPN device with the opposite type transistor whose emitter is tied directly either to ground (for the overshoot case) or to the positive power supply (for the undershoot case). For the overshoot case, the output node sources PNP emitter current, and a sufficiently large PNP collector current will turn on the NPN. They interpret latchup to occur when the PNP bypass current reverses direction. This interpretation is incorrect, however. As shown in a later section, latchup occurs when the derivative of bypass current with respect to the sourced current changes sign, not when the current changes sign.

Next to the gain condition for the parasitic transistors, the most frequently cited latchup parameter is holding current. The origin of the terminology can be traced to early SCR work and is defined as the minimum current for which, in the absence of any noise, the PNPN structure will remain forever in the low impedance state. (It is possible to transiently observe currents below the holding current when switching from the latched to the blocking state. See appendix A.) An equation for the holding current of the lumped equivalent circuit, including the effects of series emitter resistance, is proposed in [Ochoa-79] and [Estreich-80]. Their equation for zero series resistance is

$$I_h = I_{rw}\beta_{fn}(\beta_{fp} + 1) + \frac{I_{rs}\beta_{fp}(\beta_{fn} + 1)}{(\beta_{fn}\beta_{fp} - 1)}$$

$$= \frac{\alpha_{fn}I_{rw} + \alpha_{fp}I_{rs}}{\alpha_{fn} + \alpha_{fp} - 1}. \qquad (4-10)$$

Note the second version of their equation for holding current is identical to the equation for turn-off current given by (4-8). This identity is obvious once one recognizes the bypass current $I_{rw}$ is the same as the negative gate current $-I_{g1}$, and, similarly, $I_{rs} = -I_{g2}$. (Recall the previous section's discussion on the necessity of bypass current or negative gate current for providing a blocking state when $\alpha_{fn} + \alpha_{fp} \geq 1$.) However, the turn-off current for an SCR tetrode is a specific, well-defined current since the gate currents $I_{g1}$ and $I_{g2}$ are set by external circuitry. Such is not the case for the parasitic PNPN structure in which the bypass currents depend on triggering mode and are configurationally related to the corresponding emitter currents. In fact, the right hand side of (4-10) describes total current through the PNPN device at any point on the I/V characteristic before saturation occurs because it is simply a statement of Kirchoff's current law. It is not valid for the latched state, where at least one of the transistors is in saturation, nor is it valid for the portion of the negative differential resistance region where at least one of the transistors is in saturation. The last point for which it is valid is the turn-off point, at which the well/substrate bias is exactly zero in its transition from reverse to forward bias. Thus, (4-10) is not a valid equation for the holding current, which is defined by zero differential resistance and separates the latched state from the low voltage end of the negative differential resistance curve. The correct equation for holding current is somewhat more complicated and is derived for the first time in section 4.5.

Raburn also uses turn-off current to approximate holding current and derives the same equation as above [Raburn-80]. He explicitly notes that it is valid when the well/substrate junction current vanishes, but he erroneously associates it with the holding point. The turn-off point actually occurs in the negative differential resistance region and is difficult to identify experimentally.

Another problem arises in comparing experimental data with the above equation. In the form shown the bypass currents $I_{rs}$ and $I_{rw}$ are not well defined. Usually one assumes some voltage across the bypass resistance adequate to turn on the transistor, but these estimates can vary by a factor of three (from 0.3 to 0.9 volts). In addition, it is not clear that both transistors would have the same forward bias at the turn-off point. Consequently, many of the theoretical/experimental comparisons of holding current that appear in the literature involve a fair amount of "fitting," and agreement between the two is often forced.

## 4.3 Physical Analysis of Latchup: A New Direction

We now return to the parasitic PNPN structure inherent to bulk CMOS for a rigorous and comprehensive investigation of its properties. Figure 4-4 shows the cross section for an N-well technology with the PN junctions labeled for subsequent discussion. Junction J1 is the emitter/base junction of the VPNP, J2 is the N-well/substrate junction, which is usually blocking, and J3 is the emitter/base junction of the LNPN. The terminals of the PNPN structure are shown connected via general elements (indicated by the boxes) to leads which in turn connect to a power supply. Whether the power supply is represented as a Thevenin or Norton equivalent is of no concern at the moment. We first concentrate on the physical relationship among the various components.

# Latchup Models and Analyses

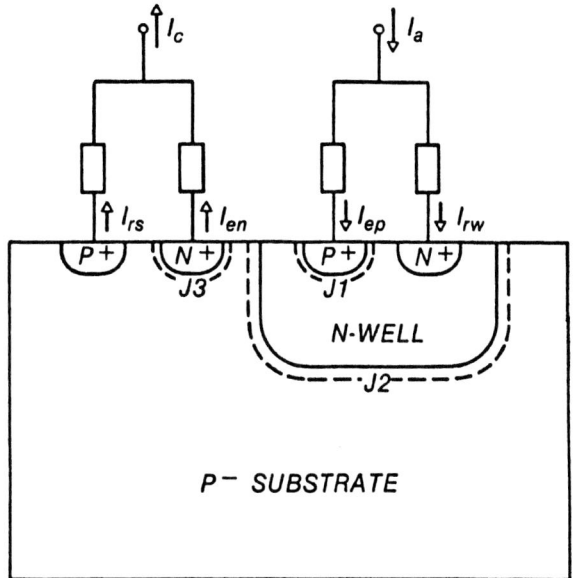

Figure 4-4. External connections for 4-terminal PNPN device. Each lead can include a current or voltage source, a resistor, an open or a short circuit.

### 4.3.1 Semiconductor Current Relationships

By way of general review we note the hole current on the substrate side of J2's space charge layer is

$$I_p = \alpha_{fp} I_{ep} + I_{pd} + I_{psc} . \qquad (4-11)$$

The first term on the right hand side accounts for holes injected through J1 into the N-well which diffuse across the N-well and are collected by J2. The second term represents holes that are thermally generated within the N-well which diffuse to J2. The third term represents holes thermally generated within the space charge layer of J2 which are swept to the space charge edge by the electric field. For simplicity the anode voltage is assumed low enough that both hole and electron multiplication factors are unity. The well/substrate breakdown voltage greatly exceeds applied voltage, except possibly when overvoltage stressing is used.

When J3 is forward biased, electrons are injected into the substrate. These electrons, plus those thermally generated in the substrate, diffuse toward the space charge layer of J2. The electrons which do not recombine in transit are swept by the space charge layer field into the N-well. The electron current at the substrate side of J2's space charge layer is

$$I_n = \alpha_{fn} I_{en} + I_{nd} . \qquad (4-12)$$

The total current flowing through the surface defined by J2's space charge edge in the substrate is then

$$I = I_n + I_p = \alpha_{fn} I_{en} + \alpha_{fp} I_{ep} + I_d + I_{sc} , \qquad (4-13)$$

where $I_d = I_{nd} + I_{pd}$ is the total diffusion current collected by J2, and $I_{sc} = I_{psc}$ is the current generated in J2's space charge region.

Before examining the above equation in more detail, we will include an important dependency between each of the emitter currents and its corresponding bypass current. The key relationship can be made rather general, and it provides a means of addressing a wide variety of latchup triggering techniques. Observe that the two relations

$$I = I_{rw} + I_{ep} \qquad (4-14)$$

and

$$I = I_{rs} + I_{en} \qquad (4-15)$$

hold for the blocking state whenever the element contained in each of the boxes shown in Figure 4-4 is either (1) a voltage source, (2) a current source, or (3) a resistor (of any finite value, including zero). In addition, the boxes on the branches marked by $I_{rs}$ and $I_{rw}$ could also be open circuits (so that $I_{rs} = 0$ or $I_{rw} = 0$).

## Latchup Models and Analyses

In the blocking state some of the total current I entering the PNPN tetrode structure bypasses the $P^+$ emitter, i.e., current through the base/emitter bypass resistor $R_w$ flows in the same direction as the emitter current for the bypassed transistor.[5] We can then define an effective large signal injection factor for the VPNP as

$$\gamma_p^* = \frac{I_{ep}}{I} = \frac{I_{ep}}{I_{ep} + I_{rw}}. \qquad (4-16)$$

The effective large signal injection factor lies in the range $0 \le \gamma_p^* \le 1$ when $I_{rw}$ flows in the same direction as $I_{ep}$. (It is defined to be 1 when $I_{rw}$ flows counter to $I_{ep}$.) Likewise, when the current exits the PNPN tetrode some of the current bypasses the $N^+$ emitter by flowing through $R_s$. The effective injection factor for the LNPN is then

$$\gamma_n^* = \frac{I_{en}}{I} = \frac{I_{en}}{I_{en} + I_{rs}}. \qquad (4-17)$$

Substituting the effective injection factors into (4-13) results in

$$I = \alpha_{fn}\gamma_n^* I + \alpha_{fp}\gamma_p^* I + I_{s2}, \qquad (4-18)$$

where $I_{s2} = I_d + I_{sc}$ is the total leakage current when the blocking junction J2 is reversed biased. Solving for the current we obtain the general result

---

[5] This direction is opposite to base current flow through $R_w$ when the LNPN is off or its emitter is left floating. It is also opposite to positive gate current flow in an SCR. Gating and bypassing have opposite effects; gating pre-disposes a PNPN device to the latched state, and bypassing pre-disposes it to the blocking state.

$$I = \frac{I_{s2}}{1 - (\alpha_{fn}\gamma_n^* + \alpha_{fp}\gamma_p^*)}, \qquad (4-19)$$

which is the starting point for a new latchup criterion to be derived in the next section.

Introducing the effective injection factor allows us to view latchup of the PNPN tetrode structure in a different manner. We can replace the tetrode structure by an equivalent diode structure with effective transistor gains $\alpha_{fn}^* = \gamma_n^*\alpha_{fn}$ and $\alpha_{fp}^* = \gamma_p^*\alpha_{fp}$. The effective alphas account for the reduced "injection efficiency" of the total current brought about by the alternate current paths through the bypass resistors. Because of the small values for $\gamma_n^*$ and $\gamma_p^*$, this equivalent diode exhibits a blocking state even when $(\alpha_{fn} + \alpha_{fp}) \geq 1$. Latchup will not occur until the effective injection factors, whose values range from 0 to 1, each reaches some critical value. The next two sections discuss the correct choice of these critical values.

### 4.3.2 Static Latchup Criterion: An Incorrect Choice

From (4-19) it appears that current will go to infinity when the denominator goes to zero, or when $(\alpha_{fn}^* + \alpha_{fp}^*) = 1$. It is tempting to view this equation as a static latchup criterion (static because the criterion depends only on the values of the relevant parameters and not on their derivatives). However, a closer inspection reveals that, strictly speaking, $(\alpha_{fn}^* + \alpha_{fp}^*) = 1$ can never be satisfied in the blocking state. This becomes clearer by rewriting (4-18) as

$$\alpha_{fn}\gamma_n^*(I) + \alpha_{fp}\gamma_p^*(I) = 1 - \frac{I_{s2}}{I}. \qquad (4-20)$$

Increasing the current I causes the right hand side to approach, but not reach, unity. Since $\gamma_n^*$ and $\gamma_p^*$ are both functions of I, they too increase, but in a way that the left hand side also approaches but does not reach unity.

Although the denominator of (4-19) can never reach zero, apparently it can come arbitrarily close, so apparently the current I can become arbitrarily large. However, as we shall see in the next section, latchup occurs well before $(\alpha_{fn}^* + \alpha_{fp}^*)$ reaches unity. The PNPN structure then switches into the low impedance state where (4-19) is no longer valid. Knowing when switching occurs is equivalent to knowing the range of stable solutions for (4-19). Thus, the correct criterion for latchup results from an examination of blocking state stability.

### 4.3.3 Differential Latchup Criterion: A Matter of Stability

To understand switching from the blocking state, it is necessary to consider the stability of any point in the blocking state when that point is perturbed by variations of any sort. Therefore, we now turn to the subject of differential stability.

The differential latchup criterion can be found by differentiating (4-19) with respect to I and rearranging the terms so that

$$\frac{dI}{dI_{s2}} = \frac{1}{1 - \frac{d[I(\alpha_{fn}^* + \alpha_{fp}^*)]}{dI}}. \qquad (4-21)$$

A specific operating point is stable if, and only if,

$$\frac{d[I(\alpha_{fn}^* + \alpha_{fp}^*)]}{dI} < 1. \qquad (4-22)$$

Otherwise, the point is unstable, and the response to a perturbation (whether deliberate or accidental), is a tendency for the current to rise uncontrollably until the transistors are driven into saturation. Consequently, the differential criterion for latchup can be written as

$$\frac{d[I(\alpha^*_{fn} + \alpha^*_{fp})]}{dI} = \alpha^*_{fn} + \alpha^*_{fp} + \frac{I\,d(\alpha^*_{fn} + \alpha^*_{fp})}{dI} \geq 1. \quad (4-23)$$

Since $(\alpha^*_{fn} + \alpha^*_{fp})$ generally increases with I, the second derivative is positive, and the differential latchup criterion is satisfied at a value of $(\alpha^*_{fn} + \alpha^*_{fp})$ less than unity.

Before applying the above differential latchup criterion, let us consider some alternative formulations. From (4-20) we note that $I(\alpha^*_{fn} + \alpha^*_{fp}) = I - I_{s2}(V_2)$, which, when substituted into (4-23) yields

$$\frac{dI_{s2}}{dI} = \left(\frac{dI_{s2}}{dV_2}\right)\left(\frac{dV_2}{dI}\right) \leq 0. \quad (4-24)$$

Since $I_{s2}$ increases with reverse bias, this expression can only hold when $dV_2/dI \leq 0$. This first version says that latchup occurs once the well/substrate reverse bias ceases to rise with increasing total current through the PNPN structure. Recall from section 4.1.2 that Gibbons used this condition as an approximation to the switching point. In fact, this condition defines the switching point, even though it occurs when dV/dI is still slightly positive, not zero.

A more convenient version involves the directly measurable terminal currents. Substituting for $\gamma^*_{ns}$ and $\gamma^*_{ps}$ from (4-16) and (4-17) into (4-18) and (4-23) yields

$$\frac{d(\alpha_{fn}I_{en})}{dI} + \frac{d(\alpha_{fp}I_{ep})}{dI} \geq 1 \quad (4-25)$$

as a second version of the differential latchup criterion. This latter version can be simplified by rewriting it in terms of small-signal alphas. Given $\alpha_{fn} = I_{cn}/I_{en}$, then

# Latchup Models and Analyses

$$I_{en}\frac{d\alpha_{fn}}{dI_{en}} = \alpha_{fns} - \alpha_{fn}, \qquad (4-26)$$

where $\alpha_{fns} = dI_{cn}/dI_{en}$ is the small-signal forward alpha for the NPN transistor. Likewise,

$$I_{ep}\frac{d\alpha_{fp}}{dI_{ep}} = \alpha_{fps} - \alpha_{fp}, \qquad (4-27)$$

where $\alpha_{fps} = dI_{cp}/dI_{ep}$ is the small-signal forward alpha for the PNP transistor. Expanding (4-25) then results in

$$\alpha_{fns}\frac{dI_{en}}{dI} + \alpha_{fps}\frac{dI_{ep}}{dI} \geq 1, \qquad (4-28)$$

which is the desired form of the second version.

Just as we defined an effective large-signal injection factor for the bypassed transistors in (4-16) and (4-17), we can define their equivalent small-signal counterparts as $\gamma_{ns}^* = dI_{en}/dI$ and $\gamma_{ps}^* = dI_{ep}/dI$. In addition, use of the chain rule on these derivatives yields the alternative forms

$$\gamma_{ns}^* = \frac{dI_{en}}{dI} = \frac{1}{1 + \dfrac{dI_{rs}}{dI_{en}}} \qquad (4-29)$$

and

$$\gamma_{ps}^* = \frac{dI_{ep}}{dI} = \frac{1}{1 + \dfrac{dI_{rw}}{dI_{ep}}}. \qquad (4-30)$$

Substituting (4-29) and (4-30) into (4-28) then produces a third version of the differential latchup criterion, viz.,

$$\frac{\alpha_{fns}}{1 + \dfrac{dI_{rs}}{dI_{en}}} + \frac{\alpha_{fps}}{1 + \dfrac{dI_{rw}}{dI_{ep}}} \geq 1. \qquad (4-31)$$

Although we have developed several specific versions for the dynamic latchup criterion given by (4-23), both (4-28) and (4-31) have the same general form, namely

$$\alpha^{*}_{fns} + \alpha^{*}_{fps} \geq 1, \qquad (4-32)$$

where the effective small-signal alphas for the bypassed transistors are determined from either of the following relationships:

$$\alpha^{*}_{fns} = \alpha_{fns}\gamma^{*}_{ns} = \frac{\alpha_{fns}}{dI/dI_{en}} = \frac{\alpha_{fns}}{1 + \dfrac{dI_{rs}}{dI_{en}}} \qquad (4-33)$$

and

$$\alpha^{*}_{fps} = \alpha_{fps}\gamma^{*}_{ps} = \frac{\alpha_{fps}}{dI/dI_{ep}} = \frac{\alpha_{fps}}{1 + \dfrac{I_{rw}}{dI_{ep}}}. \qquad (4-34)$$

For this final form we have written the derivatives assuming emitter currents to be the independent variables. In the characterization chapter we shall see that using a current source in the emitter terminal is the preferred 4-terminal characterization technique for the blocking state and that it yields measurements directly applicable to the above version of the differential latchup criterion.

# Latchup Models and Analyses 63

In summary, we see two sources of change that can destabilize a point in the blocking state. The first is an increase in actual small-signal alphas, the second, an increase in effective small signal injection factors. Most modern CMOS technologies produce a vertical parasitic bipolar with excellent low current characteristics so that, until reduced by high level injection effects, its forward alpha is well approximated by a constant as long as the PNPN current is below the switching current. The same approximation can often be made for the parasitic lateral bipolar as well, especially in epi-CMOS. Consequently, the small-signal alphas can often be replaced by large-signal alphas (whose sum usually exceeds unity even at very low currents), and it is usually the rise in effective small-signal injection factors that destabilizes the blocking state. For this latter case, once we know how total device current depends on emitter currents or how each of the bypass currents depends on its corresponding emitter current, we know the effective small-signal injection factors and can readily determine the current levels necessary for latchup.

### 4.3.4 High Level Injection Effects

All that is needed to implement the differential latchup criterion just derived is a transistor model relating the emitter and bypass currents. A small bypass resistance on one transistor may require so much current from the opposite transistor before the switching current is reached that high level injection effects become important. Accordingly, the emitter current for the parasitic NPN in the forward active region is written as

$$I_{en} = \frac{I_{eno}}{q_{bn}}, \qquad (4-35)$$

where the emitter current in the absence of high level injection is given by

$$I_{eno} = I_{sn}[e^{V_{ben}/V_t} - 1], \qquad (4-36)$$

where $I_{sn}$ is the saturation current for the base/emitter junction. The normalized high-level injection parameter is given by

$$q_{bn} = 0.5[q_{1n} + (q_{1n}^2 + 4q_{2n})^{0.5}], \qquad (4-37)$$

where the effects of basewidth modulation are included in

$$q_{1n} = 1 + \frac{V_{ben}}{|V_{Bn}|} + \frac{V_{bcn}}{|V_{An}|}, \qquad (4-38)$$

and the effects of high level injection are included in

$$q_{2n} = \frac{I_{eno}}{I_{kfn}}. \qquad (4-39)$$

The bipolar transistor parameters $V_{An}$, $V_{Bn}$, and $I_{kfn}$ have their usual meaning. $V_{An}$ is the Early voltage and $V_{Bn}$ the inverse Early voltage for the parasitic NPN. $I_{kfn}$ is the knee for the collector current on a logarithmic scale plotted vs. $V_{ben}$. (For more details, see section 2.5 of [Getreau-78].)

The appropriate differential relationship for the latchup criterion is then given by

$$\frac{dI_{en}}{dI_{rs}} = \left(\frac{1}{r_{en}}\right)\frac{dV_{be}}{dI_{rs}}, \qquad (4-40)$$

where the small-signal emitter resistance, including high level injection effects, is

$$r_{en} = \frac{dV_{ben}}{dI_{en}} \simeq \frac{V_t}{(1-H_n)I_{en}}. \qquad (4-41)$$

The new high level injection parameter is defined by

$$H_n = \frac{0.5\zeta_n}{[1 + \zeta_n + (1+\zeta_n)^{0.5}]}, \qquad (4-42)$$

where

$$\zeta_n = \frac{4q_{2n}}{q_{1n}^2} \qquad (4-43)$$

increases monotonically from zero as high level injection sets in and approaches 0.5 as an asymptotic limit when emitter current greatly exceeds the knee current. The parameter $q_{bn}$ also increases (from a value $q_{bn} = q_{1n}$ at low level injection) as high level injection sets in, and it is this term that reduces the exponential slope of $I_{en}$ vs. $V_{ben}$ from 1 to 0.5. Finally, including the effects of series resistance $R_{en}$ in the emitter leg and series resistance $R_{bn}$ in the base leg, the above derivative becomes

$$\frac{dV_{be}}{dI_{rs}} = R_s - [R_{en} + R_{bn}(1 - \alpha_{fns})]\frac{dI_{en}}{dI_{rs}}. \qquad (4-44)$$

Substituting (4-40) and (4-44) into (4-33) then yields the effective small-signal NPN alpha

$$\alpha_{fns}^* = \frac{R_s \alpha_{fns}}{R_s + r_{en} + R_{en} + R_{bn}(1 - \alpha_{fns})}. \qquad (4-45)$$

In an analagous manner (4-34) becomes

$$\alpha_{fps}^* = \frac{R_w \alpha_{fps}}{R_w + r_{ep} + R_{ep} + R_{bp}(1 - \alpha_{fps})}. \qquad (4-46)$$

In summary, high-level injection effects increase the small-signal emitter resistance, thus reducing the effective small-signal alpha and making latchup less likely. In most cases the emitter and base series resistances $R_e$ and $R_b$ are negligible compared to $r_e$ in the blocking state, and in the following sections these terms are often omitted for clarity.

## 4.4 SAFE Space: A Rigorous Definition of the Blocking State

The previous section developed several versions of the dynamic latchup criterion and noted that destabilization of the blocking state is usually caused by a rise in the effective small-signal injection factor. In this section we alter our viewpoint and use the same stability analysis to explore the conditions for latchup-free operation. Stable solutions for the blocking state occur when

$$\alpha^*_{fns} + \alpha^*_{fps} < 1, \qquad (4-47)$$

which defines a triangularly shaped region in effective small-signal alpha space, as depicted in Figure 4-5. We shall refer to the region under (but not including) the hypotenuse as SAFE space, since solutions for the blocking state lying in this region are safe from latchup. Points on the hypotenuse are switching points, so the hypotenuse forms the switching border of SAFE space.

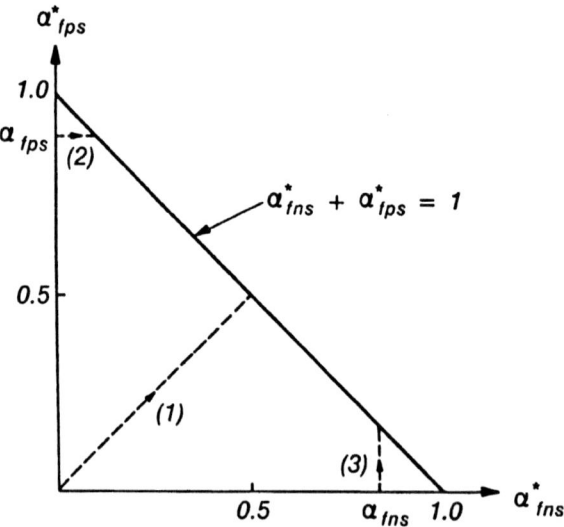

Figure 4-5. SAFE space representation of the blocking state. Latchup does not occur as long as the operating point remains in the triangularly shaped region. Dashed lines show special case operating loci: (1) identically complementary transistors; (2) floating N-well; and (3) floating substrate.

Latchup Models and Analyses

Once the entire set of elements for the PNPN lumped equivalent model is specified, the set of possible solutions (under different excitations) forms a locus in SAFE space. The complete locus describes a path from the origin to the switching border. A figure consisting of several paths, where each path corresponds to a different specification for the set of elements, is a compact way of illustrating blocking state behavior, and we will refer to such a diagram as a map of SAFE space. An appreciation of path types can be gained by considering several special cases.

### 4.4.1 Symmetric PNPN Structure

In this case the NPN and PNP transistors are identically complementary (i.e., $\alpha_{fn} = \alpha_{fp}$ and $I_{sn} = I_{sp}$) in the forward active region, and the bypass resistors are equal (i.e., $R_s = R_w$). Then, $\alpha^*_{fns} = \alpha^*_{fps}$, and stable solutions for the blocking state are constrained to lie on the dashed line at 45 degrees to the axes in Figure 4-5. The exact location of the solution point depends on the actual values for the transistor parameters and bypass resistors, but typically is very close to the origin. An increase in well/substrate junction current, as caused by radiation or by high temperatures, would move this point along the line toward the hypotenuse.

The differential latchup criterion now becomes $\alpha^*_{fns} = \alpha^*_{fps} \geq 0.5$, and with the help of equation (4-33) can be expanded to read

$$1 + \frac{dI_{rs}}{dI_{en}} \leq 2\alpha_{fns} . \qquad (4-48)$$

Using the expression derived in the previous section for the derivative, but neglecting series emitter and base resistance, then yields the condition which states to what degree the NPN must be turned on before latchup occurs, namely

$$\frac{r_{en}}{R_s} \leq (2\alpha_{fns} - 1) . \qquad (4-49)$$

Substituting from (4-41) into the above form of the differential latchup criterion then yields the emitter current at the switching point $I_{en,s}$, namely,

$$I_{en,s} = \frac{I_{en,so}}{q_{bn}} = \frac{V_t}{R_s(1-H_n)(2\alpha_{fns}-1)}, \qquad (4-50)$$

where use has been made of the fact that emitter current at the switching point generally exceeds reverse-bias leakage current by many orders of magnitude.

Bypass resistance multiplied by emitter current at the switching point is then given by

$$R_s I_{en,s} = \frac{V_t}{(1-H_n)(2\alpha_{fns}-1)}, \qquad (4-51)$$

which depends only on temperature, the transistor's small-signal alpha, and the high-level injection factor. The switching current $I_s$ itself is the sum of this emitter current and corresponding bypass current and is given by

$$I_s = I_{en,s} + \frac{V_t}{R_s}[\ln(1+\frac{q_{bn}I_{en,s}}{I_{sn}})], \qquad (4-52)$$

with the second term generally larger than the first. Strictly speaking, switching current is not inversely proportional to $R_s$, as is the emitter current at the switching point. The base/emitter bias on the NPN at the switching point is given by

$$V_{ben,s} = V_t \ln[\frac{q_{bn}V_t}{(1-H_n)(2\alpha_{fns}-1)I_{sn}R_s}], \qquad (4-53)$$

which is weakly dependent on the bypass resistance and accounts for the fact that switching current times bypass resistance is not a constant for a specific PNPN device held at a constant temperature, as has been frequently assumed by many authors.

### 4.4.2 Floating N-well Triode

Floating the N-well is equivalent to operating the LNPN with a hook collector. In this case there is no bypass resistor for the VPNP, and (4-34) simplifies to $\alpha^*_{fps} = \alpha_{fps}$. If, for simplicity, we assume $\alpha_{fps}$ to be constant over the range of interest for transistor current, stable blocking state solutions are now constrained to lie on the horizontal dashed line in Figure 4-5. The larger the value for $\alpha_{fps}$, the closer these solutions will be to the northern corner of SAFE space.

For this special case the differential latchup criterion simplifies to

$$\alpha^*_{fns} = \frac{\alpha_{fns}}{dI/dI_{en}} \geq 1 - \alpha_{fps}, \qquad (4-54)$$

which can be re-arranged to yield a condition on the incremental changes in PNPN currents as

$$\frac{dI}{dI_{en}} = 1 + \frac{dI_{rs}}{dI_{en}} \leq \frac{\alpha_{fns}}{1 - \alpha_{fps}}. \qquad (4-55)$$

Since the NPN emitter current is determined by the current through the bypass resistor (here we assume both emitter and resistor are connected directly to ground), the above condition yields a specific emitter current at which latchup occurs, namely

$$I_{en,s} = \frac{(1 - \alpha_{fps})V_t}{(\alpha_{fns} + \alpha_{fps} - 1)(1 - H_n)R_s}. \qquad (4-56)$$

Again the series base and emitter leg resistances and the saturation current $I_{sn}$ are all assumed negligible at the switching point. The base-to-emitter bias on the NPN at the switching point is then given by

$$V_{ben,s} = V_t \ln\left[\frac{q_{bn}(1 - \alpha_{fps})V_t}{(1 - H_n)(\alpha_{fns} + \alpha_{fps} - 1)I_{sn}R_s}\right], \quad (4-57)$$

and the switching current itself is given by

$$I_s = I_{en,s} + \frac{V_t}{R_s}\left[\ln\left(1 + \frac{q_{bn}I_{en,s}}{I_{sn}}\right)\right], \quad (4-58)$$

where again the second term usually dominates the first.

Again we see that the product of bypass resistance and emitter current at the switching point is a function only of temperature, both small-signal forward alphas, and the high-level injection factor. Because the critical base/emitter forward bias depends on bypass resistance, the same is not strictly true for the product of bypass resistance and switching current.

Although the derivations in this chapter have started with the concept of an effective injection efficiency, the starting viewpoint could just as well have been the parallel conductances formed by a transistor and its bypass resistor. This latter concept was used in Chapter 2 to derive a latchup stability criterion for the floating N-well. Setting $\Delta I_j/\Delta I_{bn} = 0$ (the usual case) and substituting $G_{bn} = (1 - \alpha_{fns})/r_{en}$ for the small-signal conductance, equation (2-2) becomes (4-56). These seemingly different viewpoints are really equivalent.

A viewpoint that is different, and which incorrectly predicts switching current, is that switching occurs when current through the bypass resistor reverses direction. To understand why this is so, note that when the VPNP emitter voltage is raised above the power supply, as in output overshoot for example, PNP base current flows out of the N-well contact before the parasitic NPN turns on, but after latchup, current flows into the N-well contact. In one analysis the condition of zero current at the well contact is used to define the switching point [Iizuka-81]. Since this condition is identical to a floating N-well, the condition of zero well current should agree with equation (4-56). However, in place of thermal voltage $V_t$ the zero current condition reads $V_{ben}$, the actual base/emitter forward bias. As such it is a large-signal latchup criterion, and it overestimates the bias and currents that cause latchup.

### 4.4.3 Floating Substrate Triode

Floating the substrate is equivalent to operating the VPNP with a hook collector. In this case there is no bypass resistor for the LNPN so that $dI/dI_{en} = 1$, and (4-33) reduces to $\alpha^*_{fns} = \alpha_{fns}$. If, for simplicity, we assume $\alpha_{fns}$ to be constant over the range of interest for transistor current, stable solutions for the blocking state are now constrained to lie on the vertical dashed line in Figure 4-5. The larger the value for $\alpha_{fns}$, the closer these solutions will be to the eastern corner of SAFE space.

The PNP emitter current at the switching point is then given by

$$I_{ep,s} = \frac{(1 - \alpha_{fns})V_t}{(\alpha_{fns} + \alpha_{fps} - 1)(1 - H_p)R_w}, \qquad (4-59)$$

which is similar in form to the NPN emitter current for the floating N-well case. The corresponding emitter-to-base bias is given by

$$V_{ebp,s} = V_t \ln\left[\frac{q_{bp}(1 - \alpha_{fns})V_t}{(1 - H_p)(\alpha_{fns} + \alpha_{fps} - 1)I_{sp}R_w}\right], \qquad (4-60)$$

and the switching current itself is given by

$$I_s = I_{ep,s} + \frac{V_t}{R_w}\left[\ln(1 + \frac{q_{bp}I_{ep,s}}{I_{sp}})\right]. \qquad (4-61)$$

Again we find that at the switching point the product of bypass resistance and emitter current is dependent only on temperature and transistor forward alphas and that critical emitter/base bias decreases logarithmically with increasing bypass resistance.

These equations for the floating substrate triode configuration could be used to assess latchup sensitivity in N-well CMOS when a substrate generator is used. A more accurate equivalent circuit for the generator would include a capacitor from substrate to ground. Using these equations for such predictions is somewhat conservative because the generator's capacitance must be discharged to its opposite polarity before the LNPN base/emitter can forward bias.

In the next section we shall find the principal application of the triode equations to be accurate approximations for the more general tetrode configuration.

### 4.4.4 General Tetrode

The most general case occurs when both bypass resistors are in place. The differential latchup criterion for the general tetrode case follows by substituting (4-45) and (4-46) into (4-32). Omitting the series emitter and base resistances for simplicity then yields

$$\frac{\alpha_{fns}}{1 + \frac{r_{en}}{R_s}} + \frac{\alpha_{fps}}{1 + \frac{r_{ep}}{R_w}} \geq 1, \qquad (4-62)$$

where the ratios of small-signal emitter resistance to bypass resistance are given by

$$\frac{r_{ep}}{R_w} = \frac{V_t}{(1 - H_p) R_w I_{ep}} \qquad (4-63)$$

and

$$\frac{r_{en}}{R_s} = \frac{V_t}{(1 - H_n) R_s I_{en}}. \qquad (4-64)$$

Since the maximum value for $\alpha_{fns}$ or $\alpha_{fps}$ is unity, the criterion cannot be satisfied until either $r_{ep} < R_w$ or $r_{en} < R_s$. Thus, a simple,

# Latchup Models and Analyses 73

conservative interpretation of the differential latchup stability criterion is to keep both transistors from turning on to the point that emitter current times bypass resistance for either transistor reaches the thermal voltage. We can do better than this, however. We will develop a less conservative, yet relatively simple approximation for switching current in the general tetrode PNPN configuration.

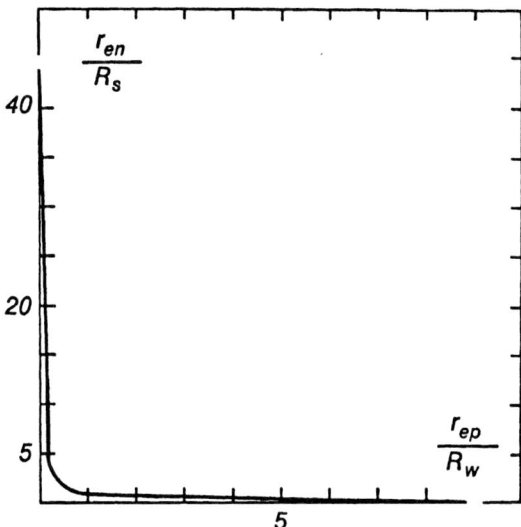

Figure 4-6. Illustration of general tetrode latchup condition. Latchup occurs in the shaded region and on its boundary. All other values of differential emitter resistance are safe. Parameter values are listed in Figure 4-7 on page 85.

Figure 4-6 depicts a graphical interpretation of the above criterion. In general, operating point trajectories through the blocking state can be represented in the quadrant pictured by points moving toward the origin. In the triode approximation, trajectories are restricted to one or the other of the two axes. The shaded area indicates the region in which latchup occurs. It is hyperbolically bounded by the equation

$$[\frac{r_{ep}}{R_w} + (1 - \alpha_{fns})][\frac{r_{en}}{R_s} + (1 - \alpha_{fps})] = \alpha_{fns}\alpha_{fps}, \qquad (4-65)$$

which defines the non-linear relationship between the emitter currents at the switching point and can be written in abbreviated form as $G(I_{en,s}, I_{ep,s}) = 0$.

Since emitter current and bypass current are related, (4-14) can be rewritten in terms of the switching current in functional form as

$$I_s = f_p(I_{ep,s}) = I_{ep,s} + \frac{V_t}{R_w}[\ln(\frac{q_{bp}I_{ep,s}}{I_{sp}})]. \qquad (4-66)$$

Similarly, (4-15) can be rewritten as

$$I_s = f_n(I_{en,s}) = I_{en,s} + \frac{V_t}{R_s}[\ln(\frac{q_{bn}I_{en,s}}{I_{sn}})]. \qquad (4-67)$$

We then have a second non-linear equation involving the two emitter currents at the switching point, namely, $F(I_{en,s}, I_{ep,s}) = f_n - f_p = 0$. The two non-linear equations can be solved iteratively to find $I_{en,s}$ and $I_{ep,s}$. Once the emitter currents are known, the corresponding base/emitter bias and switching current can be found as discussed above.

Although iteratively solving these two simultaneous, non-linear equations is not difficult, such an approach would not immediately provide much insight, and for most cases of interest the triode approximation proves more than adequate. The differential latchup criterion for a tetrode given by (4-62) can be replaced by that for a triode if in the blocking state one transistor turns on well before the other. To make this idea more tangible, consider what conditions are necessary for the floating substrate triode approximation to be valid.

Note from (4-62) that $R_s$ is infinite for a floating substrate, and the first term is replaced by $\alpha_{fns}$. This replacement is also a reasonable approximation in the general case of finite $R_s$ as long as $r_{en} << R_s$, which obtains when the NPN is turned on hard enough. To elaborate, consider the NPN to be turned on

## Latchup Models and Analyses

sufficiently hard that $R_s = Kr_{en}$. The parameter $K > 1$ provides a handle on the accuracy of the triode approximation because the error introduced by replacing the first term in (4-62) with $\alpha_{fns}$ is $1/(K + 1)$. The parameter value $K = 9$ corresponds to 10% accuracy, which is a reasonable number since the current gain often is not known any more accurately.

The condition that the PNPN still be in the blocking state when the above condition holds is given by

$$r_{ep} > \frac{R_w(\alpha_{fns} + \alpha_{fps} - 1)}{(1 - \alpha_{fns})}, \qquad (4-68)$$

which can also be written in terms of the PNP emitter/base voltage as

$$V_{ebp} < V_t \left[ \frac{(1 - \alpha_{fns})V_t}{(\alpha_{fns} + \alpha_{fps} - 1)I_{sp}R_w} \right]. \qquad (4-69)$$

In addition, because PNP emitter current is still quite low, PNP bypass current can be written as

$$I_{rw} \simeq I_{rs} + I_{en} = \frac{V_{ben} + KV_t}{R_s}. \qquad (4-70)$$

Neglecting series base and emitter resistance, the PNP emitter/base voltage can then be expressed as,

$$V_{ebp} = I_{rw}R_w \simeq \frac{R_w(V_{ben} + KV_t)}{R_s}. \qquad (4-71)$$

Substituting for $V_{ben}$ from (4-35) and (4-36) and then substituting into (4-69) yields the following condition for the bypass resistors:

$$\frac{R_w}{R_s} < \frac{\ln\left[\frac{(1-\alpha_{fns})V_t}{(\alpha_{fns}+\alpha_{fps}-1)R_w I_{sp}}\right]}{K+\ln\left(\frac{KV_t}{R_s I_{sn}}\right)}. \qquad (4-72)$$

Thus, as long as the bypass resistors satisfy the above inequality, the floating substrate approximation is valid. Because the saturation currents $I_{sn}$ and $I_{sp}$ usually are extremely small, the right hand side can often be as high as 1/2 for K = 9. The analagous inequality for the bypass resistors that insures the floating N-well triode approximation is valid can be found from (4-72) by interchanging the "w" and "s" subscripts (except on the saturation currents) and the "n" and "p" subscripts. The general tetrode approximation is needed only when the bypass resistors are nearly equal.

Section 4.6.2 illustrates how the triode approximation to the tetrode configuration can provide solutions to well within 20% (usually, only a few per cent) of a numerical solution to the two simultaneous non-linear equations. The resulting equation for switching current is simple and depends on a few, easily determined parameters (none of which are "fitting" parameters).

## 4.5 Saturation Region Modeling: A New View

For several reasons, saturation region modeling of the PNPN device is more difficult than blocking state modeling. In saturation the cooperative interaction between the two parasitic bipolars can drastically alter the patterns of current flow occurring in the blocking state; forward alphas measured in the usual manner may not exemplify the saturation region. In addition, the reverse transistor action must be included in the analysis once the device current exceeds the turn-off current.

Nonetheless, much latchup work has focused on the holding point, and several prevention schemes suggest raising holding current (or holding voltage) above that capable of being provided by the power supply. Unfortunately, the equation for holding current appearing in the literature, and given by (4-10), is invalid at the holding point. This section derives the correct equation for the case $R_{s2} = R_{w2} = 0$ and discusses the difference between it and the previously presented equation.

### 4.5.1 Current Equations

When the resistors $R_{s2}$ and $R_{w2}$ in Figure 4-3 on page 47 are set equal to zero, any forward bias $V_r$ on the well/substrate junction occurs over the entire junction and results in a reverse transistor current, which is described here by $I_r = I_{s2}[\exp(V_r/V_t) - 1]$. Accounting for the reverse emitter current, as well as bypass current, the total anode current is

$$I_a = I_{rw} + I_{ep} - \alpha_{rp}I_r, \qquad (4-73)$$

where $\alpha_{rp}$ is the reverse common-base current gain of the parasitic PNP. Likewise, the cathode current is

$$I_c = I_{rs} + I_{en} - \alpha_{rn}I_r, \qquad (4-74)$$

where $\alpha_{rn}$ is the reverse common-base current gain of the parasitic NPN. In a similar fashion the current at the well/substrate junction can be written as

$$I_{wj} = \alpha_{fn}I_{en} + \alpha_{fp}I_{ep} - I_r. \qquad (4-75)$$

Finally, current continuity dictates that $I_a = I_{wj} = I_c = I$.

Manipulating the above three equations to eliminate the emitter currents results in the expression

$$(\alpha_{fn} + \alpha_{fp} - 1)I = (1 - \alpha_{fn}\alpha_{rn} - \alpha_{fp}\alpha_{rp})I_r$$

$$+ \alpha_{fn}I_{rs} + \alpha_{fp}I_{rw}, \qquad (4-76)$$

which is a general relationship among total device current I, reverse emitter current $I_r$, and bypass currents $I_{rs}$ and $I_{rw}$. A notable special case exists for $V_r = 0$. Then $I_r = 0$, and the corresponding device current,

$$I = I_{to} = \frac{\alpha_{fn} I_{rs,\,to} + \alpha_{fp} I_{rw,\,to}}{\alpha_{fn} + \alpha_{fp} - 1}, \qquad (4-77)$$

is called the turn-off current. The bypass currents at the turn-off point, $I_{rs,\,to}$ and $I_{rw,\,to}$, are specific currents and are known once the specific emitter forward biases are known at the turn-off point.

When $V_r < 0$, the term involving $I_r$ is negative and negligibly small for most cases of interest. Neglecting this term results in an expression for I that many authors have called the holding current. In fact, such an expression is generally valid for any current $I \leq I_{to}$, depending on the values of bypass current (or emitter forward bias), but it is not valid at the holding point since $I_h > I_{to}$.

To continue with the holding current derivation, we note that once $V_r > 0$, the bypass currents change little as I increases. Any further increase in I, to a good approximation, shows up only as an increase in emitter currents. Thus, when the current $I > I_{to}$, (4-76) can be approximated as

$$I \simeq I_{to} + \frac{(1 - \alpha_{fn}\alpha_{rn} - \alpha_{fp}\alpha_{rp}) I_r}{\alpha_{fn} + \alpha_{fp} - 1}. \qquad (4-78)$$

This equation relates increases in I to increases in the reverse emitter current. Bypass current increases little in saturation, and the additional device current is forward emitter current, which is balanced by the reverse transistor current.

Another important property follows from the assumption of fixed bypass currents when $I > I_{to}$. This assumption allows the original set of non-linear current equations to be linearized. Then the emitter currents can be shown to rise linearly with PNPN device current. The explicit equations are as follows:

$$(1 - \alpha_{fn}\alpha_{rn} - \alpha_{fp}\alpha_{rp})I_{ep} = C_1 I - C_2 \qquad (4-79)$$

and

$$(1 - \alpha_{fn}\alpha_{rn} - \alpha_{fp}\alpha_{rp})I_{en} = C_3 I - C_4, \qquad (4-80)$$

where

$$C_1 = (1 - \alpha_{fn}\alpha_{rn}) - \alpha_{rp}(1 - \alpha_{fn}), \qquad (4-81a)$$

$$C_2 = (1 - \alpha_{fn}\alpha_{rn})I_{rw,to} + \alpha_{fn}\alpha_{rp}I_{rs,to}, \qquad (4-81b)$$

$$C_3 = (1 - \alpha_{fp}\alpha_{rp}) - \alpha_{rn}(1 - \alpha_{fp}), \qquad (4-81c)$$

and

$$C_4 = (1 - \alpha_{fp}\alpha_{rp})I_{rs,to} + \alpha_{fp}\alpha_{rn}I_{rw,to}. \qquad (4-81d)$$

Note that when the reverse alphas are set equal to zero, these equations assume the particularly simple form $I_{ep} = I - I_{rw,to}$ and $I_{en} = I - I_{rs,to}$, which says that up to the turn-off point device current flows through the bypass resistors, and device current exceeding the turn-off current flows via emitter currents. These linearized equations will be used later to find the relationship between the emitter currents at the holding point and the holding current.

Next we elaborate on the reverse emitter current in (4-78). The specific reverse current required is that occurring at the holding point, and its formula can be found from the condition for zero differential resistance.

### 4.5.2 Differential Resistance

Total voltage across the PNPN device is given by $V = V_{ben} + V_{ebp} - V_r$, where $V_{ben}$ and $V_{ebp}$ are the emitter forward biases across the NPN and PNP, respectively. Differentiating with respect to I then yields the following expression for differential resistance:

$$(1 - \alpha_{rns}\alpha_{fns}^* - \alpha_{rps}\alpha_{fps}^*)\frac{dV}{dI} = (R_{nq} + R_{pq})(1 - \alpha_{rns}\alpha_{fns}^* - \alpha_{rps}\alpha_{fps}^*)$$

$$- (r_r - \alpha_{rps}R_{pq} - \alpha_{rns}R_{nq})(\alpha_{fns}^* + \alpha_{fps}^* - 1) . \qquad (4-82)$$

The term $R_{nq} = r_{en}R_s/(r_{en} + R_s)$ is the parallel resistance of the small-signal NPN emitter resistance and NPN bypass resistance. Similarly, $R_{pq} = r_{ep}R_w/(r_{ep} + R_w)$ is the parallel resistance for the PNP. The term $r_r = V_t/(I_r + I_{s2})$ is the small-signal reverse emitter resistance. Note that the forward and reverse alphas are now small-signal values, which is a consequence of the differentiation. The above equation has general validity for any region of the PNPN current/voltage characteristic. An examination of its properties in the various regions will be helpful before continuing the holding current derivation.

In the blocking state $r_r$ is very large, and $(\alpha_{fns}^* + \alpha_{fps}^*) < 1$. The differential resistance between the anode and cathode of the PNPN is then given by

$$\frac{dV}{dI} \simeq \frac{r_r[1 - (\alpha_{fns}^* + \alpha_{fps}^*)]}{1 - \alpha_{rns}\alpha_{fns}^* - \alpha_{rps}\alpha_{fps}^*} . \qquad (4-83)$$

As $(\alpha_{fns}^* + \alpha_{fps}^*)$ approaches unity, the differential resistance begins to decrease.

At the switching point (defined by $dV_r/dI = 0$), the sum $(\alpha_{fns}^* + \alpha_{fps}^*) = 1$, so that the differential resistance becomes

## Latchup Models and Analyses

$$\frac{dV}{dI} = \frac{dV_{ben}}{dI} + \frac{dV_{ebp}}{dI} = R_{nq} + R_{pq}. \qquad (4-84)$$

As noted previously, switching from the blocking state occurs for a slightly positive differential resistance, not zero, although it is often difficult to experimentally distinguish between the two points.

At the holding point (defined by $dV/dI = 0$ and $I > I_{to}$), $r_r$ is no longer extremely large since the PNPN is in saturation. Setting $dV/dI = 0$ in (4-82) then yields as a condition for the small-signal saturation resistance at the holding point,

$$r_{r,h} = C_n R_{nq,h} + C_p R_{pq,h}, \qquad (4-85)$$

where the coefficients

$$C_n = \frac{(1 - \alpha^*_{fps}\alpha_{rps}) - \alpha_{rns}(1 - \alpha^*_{fps})}{(\alpha^*_{fns} + \alpha^*_{fps} - 1)} \qquad (4-86)$$

and

$$C_p = \frac{(1 - \alpha^*_{fns}\alpha_{rns}) - \alpha_{rps}(1 - \alpha^*_{fns})}{(\alpha^*_{fns} + \alpha^*_{fps} - 1)} \qquad (4-87)$$

are to be evaluated at the holding point. Finally, reverse emitter current at the holding point can be written as $I_{r,h} = V_t/r_{r,h}$.

### 4.5.3 Holding Current

Before assembling the relevant equations and completing the holding current derivation, a few approximations can conveniently simplify the algebra. When the parasitic transistors are both heavily saturated, as they are at the holding point, each small-signal emitter resistance is much smaller than its corresponding bypass resistance. Consequently, at the holding point, $R_{nq,h} \simeq r_{en,h} = V_t/I_{en,h}$ and $R_{pq,h} \simeq r_{ep,h} = V_t/I_{ep,h}$. Secondly, the effective small-signal alphas simplify so that $\alpha^*_{fns} \simeq \alpha_{fns}$ and $\alpha^*_{fps} \simeq \alpha_{fps}$. Thirdly, the original current equations involve large-signal alphas, and the condition $dV/dI = 0$ leading to the reverse emitter current at the holding point introduced small-signal alphas. Strictly speaking, the holding current equation then contains both. For simplicity, this derivation assumes $\alpha_{fns} = \alpha_{fn}$ and $\alpha_{fps} = \alpha_{fp}$.

The holding current derivation now proceeds by substituting (4-79) and (4-80) into (4-85), then solving for $I_{r,h}$ and substituting it into (4-78). After some tedious but straightforward algebraic manipulations, one arrives at the following equation for holding current:

$$I_h \simeq I_{to} + (I_{to}^2 - I_m^2)^{0.5}, \qquad (4-88)$$

where

$$I_m^2 = \frac{(M_1 I_{rs,to}^2 + M_2 I_{rw,to}^2 + M_3 I_{rs,to} I_{rw,to})}{D(\alpha_{fn} + \alpha_{fp} - 1)}, \qquad (4-89)$$

$$M_1 = \alpha_{fn}\{(1 - \alpha_{fp}\alpha_{rp})[(1 - \alpha_{fn}\alpha_{rn}) + \alpha_{rp}(\alpha_{fn} - \alpha_{fp})]$$

$$- \alpha_{fn}\alpha_{rn}\alpha_{rp}(1 - \alpha_{fp})\}, \qquad (4-90a)$$

$$M_2 = \alpha_{fp}\{(1 - \alpha_{fn}\alpha_{rn})[(1 - \alpha_{fp}\alpha_{rp}) + \alpha_{rn}(\alpha_{fp} - \alpha_{fn})]$$

$$- \alpha_{fp}\alpha_{rp}\alpha_{rn}(1 - \alpha_{fn})\}, \qquad (4-90b)$$

$$M_3 = 1 - \alpha_{fn}\alpha_{rn} - \alpha_{fp}\alpha_{rp} + (1 - \alpha_{fp}\alpha_{rp})(2\alpha_{fn} - 1)\alpha_{fp}\alpha_{rp}$$

$$+ (1 - \alpha_{fn}\alpha_{rn})(2\alpha_{fp} - 1)\alpha_{fn}\alpha_{rn}, \qquad (4-90c)$$

# Latchup Models and Analyses

and

$$D = [(1 - \alpha_{fn}\alpha_{rn}) - \alpha_{rp}(1 - \alpha_{fn})][(1 - \alpha_{fp}\alpha_{rp}) - \alpha_{rn}(1 - \alpha_{fp})].$$

(4 − 90d)

Thus, through $I_m^2$ the holding current depends on the reverse, as well as forward, alphas. Even when $\alpha_{rn} = \alpha_{rp} = 0$, $I_m^2$ does not vanish, although it greatly simplifies. Since the bypass currents at the turn-off point are inversely proportional to the bypass resistances, so is the holding current, although in a somewhat more complicated manner than suggested by $I_{to}$ alone. Consequently, design strategies to minimize bypass resistance do, in fact, raise the holding current as well as the turn-off current. Finally, note that (4-88) is really an approximation to the holding current. The equation for holding current is not as precise as that for switching current, and it is more complicated. It is made even more complicated by the presence of non-zero values for $R_{s2}$ and/or $R_{w2}$.

### 4.5.4 Holding Voltage

Although a general derivation of holding voltage for the lumped element model is beyond the scope of this work, some approximation of its dependence on the relevant parameter values can be gained by considering the case of $R_{w2} = 0$ but $R_{s2} \neq 0$. This is a reasonable approximation to the usual 4-terminal PNPN test structures used for characterization. It is further assumed that $R_{s2}$ is large enough so that at the holding point the VPNP is in saturation, but the LNPN is still in the forward active mode, held there by the ohmic drop across $R_{s2}$. Subject to the above approximations, the holding voltage is easily shown to be

$$V_h = (V_{ben} + V_{ebp} - V_r)$$
$$+ R_{s2}[(1 - \alpha_{fn})I_h + \frac{\alpha_{fn}V_{ben}}{R_{s1}}], \qquad (4-91)$$

where all the quantities have been defined previously and are to be evaluated at the holding point.

The junction voltages $V_{ben}$, $V_{ebp}$, and $V_r$ are all less than 1 volt, so that when $R_{s2} = 0$, the holding voltage is typically less than 1 volt. For non-zero $R_{s2}$, however, holding voltage can be raised by (1) increasing $R_{s2}$, (2) decreasing $R_{s1}$, and (3) increasing $I_h$, which is also accomplished by decreasing $R_{s1}$. Resistance $R_{s2}$ increases with $N^+/P^+$ spacing at a fixed epitaxial layer thickness. Resistance $R_{s1}$ decreases with decreasing epitaxial layer flat-zone thickness (at a fixed epi/substrate transition zone thickness), and with decreasing transition zone thickness (at a fixed flat zone thickness). Two-dimensional numerical simulations have indeed shown holding voltage to increase accordingly [Taur-84].

## 4.6 Illustration of Latchup: Type 2 Triggering

Previous sections have developed large-signal and differential latchup criteria, and it is now time to discuss their interpretation and application in more detail. We will use a specific example first to compare predictions made by the two criteria and then to investigate the relationship between the switching point and parameters for the lumped element model.

### *4.6.1 A Sample Analysis*
For this example we add a well-to-substrate current source to the lumped element model discussed earlier. This current source could be considered representative of avalanche current when the reverse-biased junction approaches breakdown. However, we are not so much interested in the detailed breakdown characteristics as we are in a simple, controlled manner for moving through SAFE space to the switching boundary.

This analysis uses an Ebers-Moll model (without high-level injection effects) for the transistors in the lumped element circuit. The transistor model is deliberately kept simple so as not to obscure the essential features of the latchup criteria. Each transistor's alpha is assumed constant over emitter current so that small-signal alpha is equal to large-signal alpha. A full set of parameter values for the PNPN lumped equivalent model is given in Figure 4-7.

| Parameter | NPN | PNP |
|---|---|---|
| $\alpha_f$ | 0.90 | 0.98 |
| $\alpha_r$ | 0.20 | 0.11 |
| $I_{sf}$ | 1.0 pA | 0.91 pA |
| $I_{sr}$ | 4.5 pA | 4.5 pA |
| Bypass R | $R_s = 1$ k$\Omega$ | $R_w = 0.25$ k$\Omega$ |

Figure 4-7. Parameter Values for Lumped Equivalent Model

A complete numerical solution was generated for each value $I_o$ of the current source. For low values of $I_o$ the transistors are off, and the current $I_o$ flows through the bypass resistors. As can be seen from Figure 4-8, $I_{rw} = I_{rs} = I_o$ until the current source exceeds 0.40 mA. The voltage drop across $R_s$ is then sufficient to produce a noticeable (on this scale) NPN emitter current, and further increases in $I_o$ produce a rapid rise in the NPN emitter current $I_{en}$. At a current source value of approximately 0.56 mA, the NPN emitter current $I_{en}$ exceeds its bypass current $I_{rs}$. From then on $I_{rs}$ increases very little since the base/emitter resistance has fallen below the bypass resistance $R_s$. The rapid rise in current flowing to ground is accompanied by a corresponding rise in current flowing from the power supply. Since $R_w$ is significantly less than $R_s$ in this example, the PNP emitter current is still small, and the rise in power supply current is seen primarily in the PNP bypass current $I_{rw}$. (Recall that $I = I_{en} + I_{rs} = I_{ep} + I_{rw}$.) At a current source value of $I_o = 0.6330441$ mA (all these calculations are carried out to seven significant digits), the device switches to the low impedance state, about which we will have more to say later.

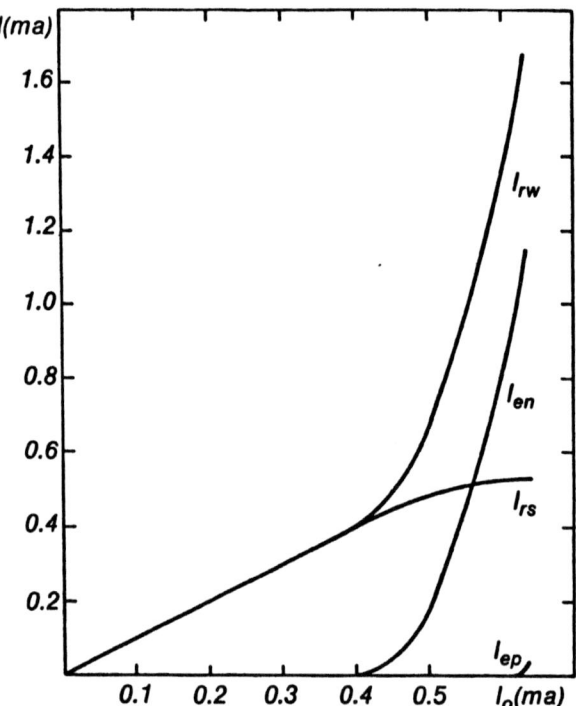

**Figure 4-8.** PNPN turn-on characteristics in blocking state. See Figure 4-7 for parameters.

Figure 4-9 compares effective large-signal and small-signal alpha sums $(\alpha_{fn}^* + \alpha_{fp}^*)$ and $(\alpha_{fns}^* + \alpha_{fps}^*)$ as a function of $I_o$. The large-signal latchup criterion is $(\alpha_{fn}^* + \alpha_{fp}^*) = 1$, and the differential latchup criterion is $(\alpha_{fns}^* + \alpha_{fps}^*) = 1$. As discussed in section 4.3.3, the latter is expected to occur first as $I_o$ is increased. That it in fact does is clear from Figure 4-9. For reference the total current I through the PNPN structure is also plotted in the figure. It is essentially identical to the plot of $I_{rw}$ in Figure 4-8 since the PNP emitter current is so small in the blocking state. The last point plotted for I is the switching current $I_s = 1.695827$ mA.

# Latchup Models and Analyses

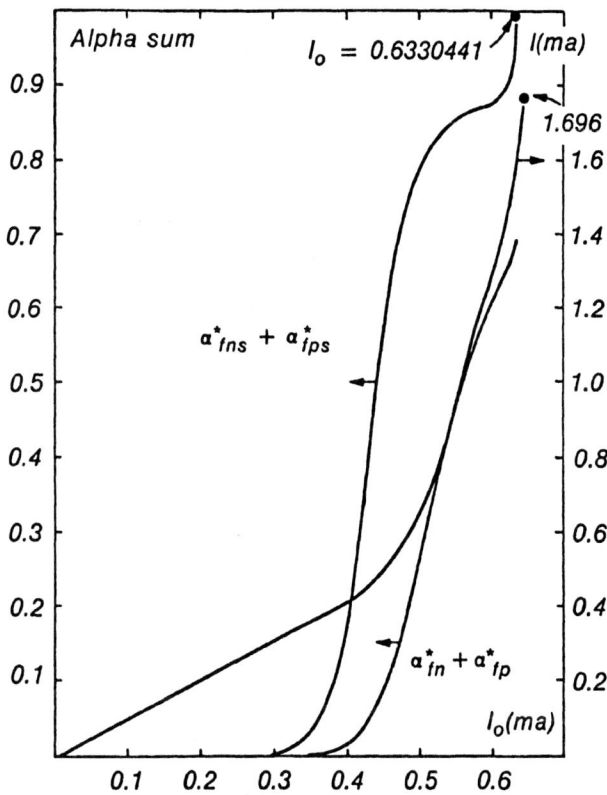

Figure 4-9. Comparison of static and differential switching criteria. The total PNPN terminal current is also shown. See Figure 4-7 for parameters.

Once the switching point is reached, the structure switches into the low impedance state, where it remains even if the current source is removed. (Note that switching is followed by a drop in voltage across the structure so that avalanche current would drop precipitously.) Figure 4-10 shows the resulting characteristic once the transistors saturate. Forward voltage on the collector/base junctions is indicated by the open circles along the characteristic. Differential resistance from anode to cathode of the PNPN device passes through zero (dV/dI=0) at a forward bias $V_r = 0.4772851$ V, and the corresponding current through the structure (holding current $I_h$) is 4.26 mA. The turn-off current (current corresponding to zero well/substrate bias) is 2.89 mA.

Note that holding current is 1.47 times the turn-off current and 2.51 times the switching current.

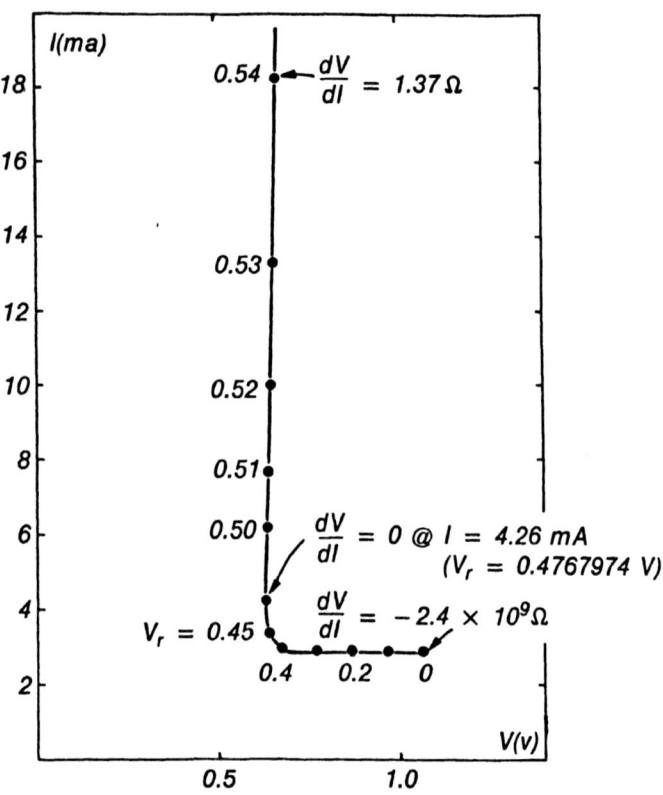

Figure 4-10.  Saturation I/V characteristics for a PNPN structure. Holding current $I_h$ = 4.26 mA, and holding voltage $V_h$ = 0.633 V at $V_r$ = 0.477 V. See Figure 4-7 on page 85 for parameters.

Figure 4-11 depicts the behavior of the four terminal currents once the transistors reach saturation. There is little increase in the bypass current since both transistors are now turned on hard. Emitter currents now rise nearly linearly with terminal current I. Note that the non-zero current difference $(I_{rw} + I_{ep}) - I$ results from reverse transistor action in the VPNP and that the non-zero current difference $(I_{rs} + I_{en}) - I$ results from reverse transistor action of the LNPN. The larger the reverse alphas, the larger these differences.

## Latchup Models and Analyses

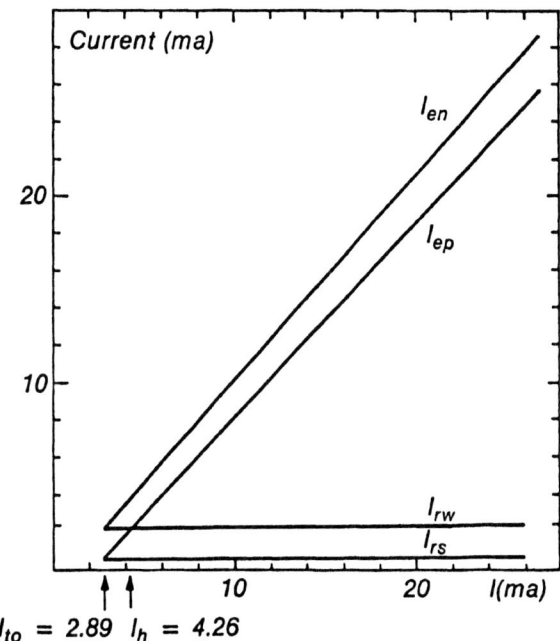

Figure 4-11. PNPN saturation bypass and emitter currents vs. terminal current. See Figure 4-7 on page 85 for parameters.

The same program also calculates $dV/dI$, the differential resistance at any point, and Figure 4-12 shows $dV/dI$ as a function of terminal current I. Note the very sharp transition from a large positive to a large negative value at a switching current $I_s = 1.695827$ mA, which is precisely the point at which $(\alpha^*_{fns} + \alpha^*_{fps}) = 1$. The corresponding sum of effective static alphas is then $(\alpha^*_{fn} + \alpha^*_{fp}) = 0.6956021$.

At low current the PNPN structure is in the blocking state, and the differential resistance exceeds 1 GΩ. As the switching current of $I_s = 1.695827$ mA is approached, the resistance drops sharply to zero (which cannot be shown on the logarithmic scale). Just above the switching current the differential resistance has a large, negative value, which drops rapidly to zero as the holding current is approached. Since the differential resistance in the latched state is less than 10 Ω, plotting the quantity $10 - dV/dI$

exhibits the entire characteristic in the vicinity of the holding point.

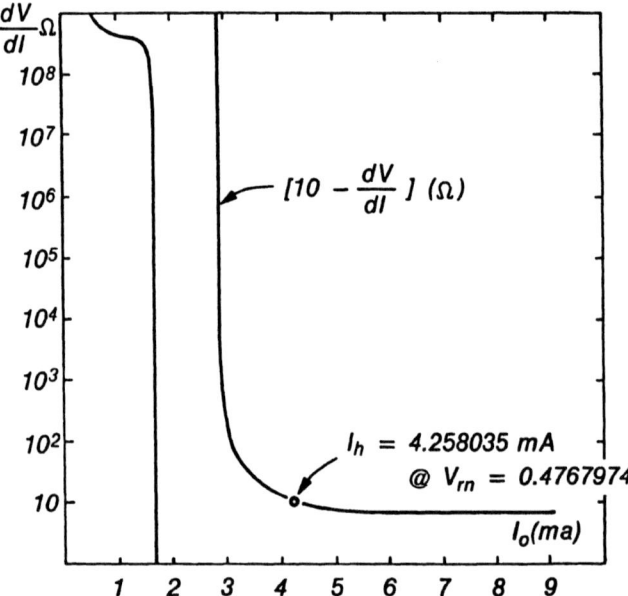

Figure 4-12. PNPN differential resistance vs. terminal current. Note the sharp reduction at the switching current ($I_s = 1.696$mA) followed by a polarity reversal. See Figure 4-7 on page 85 for parameter values.

### 4.6.2 A SAFE Space Map and Switching Current

In the previous section we analyzed the lumped element model for a specific set of parameter values. In this section we explore the behavior of the lumped element model over a wide range of parameter values. Of particular interest is how the properties of the blocking state change as the switching point is approached.

Our first view of these properties is the map of Figure 4-13 showing the various paths traversed in the effective small-signal gain space as $I_o$ is increased. In general, for each specific set of parameter values (such as $R_w$, $R_s$, $\alpha_{fn}$, $\alpha_{fp}$, $I_{sn}$, $I_{sp}$, etc.) there is a unique path from the origin of SAFE space to the switching boundary defined by the differential latchup criterion. For the case at hand each path represents the effective small-signal gain values $\alpha^*_{fns}$ and $\alpha^*_{fps}$ for a different value of well resistance $R_w$.

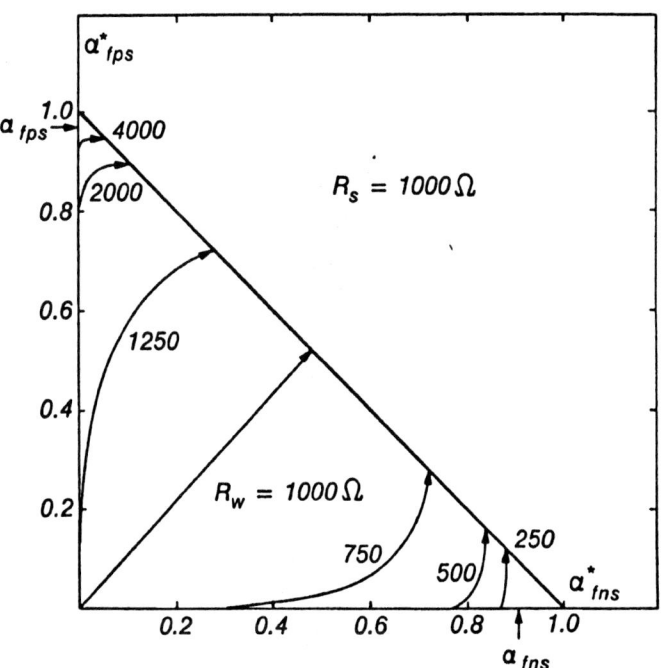

Figure 4-13. A representative map of SAFE space. Increasing N-well/substrate current moves the operating point in SAFE space from the origin to the switching boundary.

Note that once the bypass resistors are unequal, the paths move to the corners of SAFE space. In fact, the path for a ratio $R_w/R_s = 4$ is already approaching the asymptotic path for a floating N-well ($I_{rw} = 0$) near the northern corner. Likewise the path for a ratio $R_s/R_w = 4$ is approaching the asymptotic path of a floating substrate ($I_{rs} = 0$) near the eastern corner. This observation suggests a simple method for finding the switching current in the general tetrode case. The iterative procedure outlined in section 4.5.4 is tedious and can be circumvented if the two bypass resistors differ by a sufficient amount. For concreteness, let us assume $R_s > R_w$. We then estimate the switching current assuming a floating substrate ($R_s = \infty$). The explicit steps are as follows:

1. Calculate the emitter current at the switching point using

$$I_{ep,so} = \frac{(1 - \alpha_{fns})V_t}{(\alpha_{fns} + \alpha_{fps} - 1)(1 - H_p)R_w}. \qquad (4-92)$$

2. Calculate the corresponding switching current using

$$I_{so} = I_{ep,so} + \frac{V_t}{R_w}\left[\ln(1 + \frac{q_{bp}I_{ep,so}}{I_{sp}})\right]. \qquad (4-93)$$

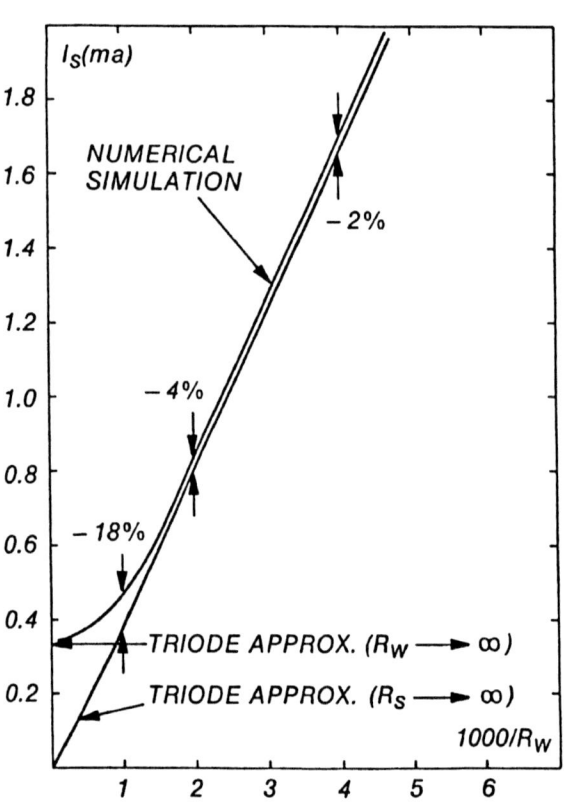

Figure 4-14. A comparison of triode and tetrode switching currents.

# Latchup Models and Analyses

Figure 4-14 compares the approximate switching current $I_{so}$ with that calculated using the numerical model. For a ratio of $R_s/R_w = 4$ the error is -1.5%, while at $R_s/R_w = 2$ the error has grown to -5%. (The floating substrate approximation slightly underestimates actual switching current.) At $R_s/R_w = 1$ it is -21%, and this latter value is the worst possible error, since for $R_s/R_w < 1$ one should use a floating N-well approximation. The open circle at $R_s = R_w$ has been calculated using (4-52), which is the equation for the symmetric PNPN structure.

Clearly, the two triode configurations can be used to find quickly a lower bound on the switching current. One needs to know only small-signal gains for the two transistors, their reverse saturation currents, and their bypass resistor values. If one bypass resistor is known to exceed the other for a particular latchup path, only the reverse saturation current and bypass resistance for the smaller one are needed. Then, removing the larger value from the lumped element model yields the appropriate triode configuration.

## 4.7 Modifications to Lumped Element Model: A Useful Perspective

Although the lumped element model has proved useful for understanding latchup behavior, determining the lumped equivalent parameters can be difficult. This is especially true for the case of the bypass resistor of the lateral transistor because of the complicated current paths through the substrate. Both majority and minority carriers can spread deeply into the substrate. For non-epi CMOS, majority carrier flow is accompanied by an electric field that can substantially influence minority carrier flow. In epi-CMOS the high/low junction formed by the epi/substrate doping profile separates the majority and minority carriers, pulling majority carriers into the highly doped substrate and forcing minority carriers toward the surface. In either case it is difficult to identify the lumped resistance that causes the lateral transistor to turn on when sufficient current passes through it.

Current flow in one portion of the substrate can cause potential variations in other, distant portions of the substrate even though most of the current does not pass through this distant portion. Very complicated resistance ladder networks have sometimes been used to account for such an effect (see, for example, [Wieder-83] or [Hall-85]), but the basic problem is that one really needs a "transfer" resistance to simply and accurately predict voltage drops in the substrate.

Transfer resistance has been calculated for a square current injector at the surface using a Fourier series [Terrill-84]. This resistance is defined as the surface potential at a point outside the injector divided by the injector current. Current is removed by a backside substrate contact, and both a lightly doped substrate and a heavily doped substrate with an epitaxial layer have been modeled. Transfer resistance is shown to decrease both with injector side length and with distance from the injector, and both effects have been experimentally verified. However, the geometries pertinent to a PNPN structure are more complicated and usually are not amenable to Fourier series analysis.

### 4.7.1 Transmission Line Model of Latchup

Troutman and Hargrove have shown that a lossy transmission line model works very well for calculating substrate potential variations prior to substantial minority carrier injection into the substrate [Troutman-83b] and [Troutman-86]. This situation describes the blocking state when the parasitic vertical bipolar turns on first. The lumped resistors normally used to model substrate resistances are then replaced by a lossy transmission line. Such a model is readily adaptable to a wide variety of contact and layout configurations, and calculating the corresponding surface potential or current distribution is then a simple transmission line problem. Dependence on distance enters the problem in a natural manner, as does whether a topside or backside contact is used. The transmission line parameters are easily related to processing parameters, such as epitaxial layer thickness, field implant and substrate out-diffusion profiles. This enables a more natural and more accurate description of the network interconnecting the two parasitic bipolars.

## Latchup Models and Analyses 95

In addition to providing general potential and current distributions in the substrate, the transmission line model simplifies calculations for a variety of two dimensional substrate resistances. Results from two-dimensional simulations agree with predicted transmission line behavior. The highly doped field region and the even more conductive substrate below the epitaxial layer both carry significant portions of the current, especially as the distance between the two topside contacts increases beyond the characteristic length of the transmission line. Even when current is removed via a backside contact, transmission line effects are important, especially if additional topside diffusions are close by.

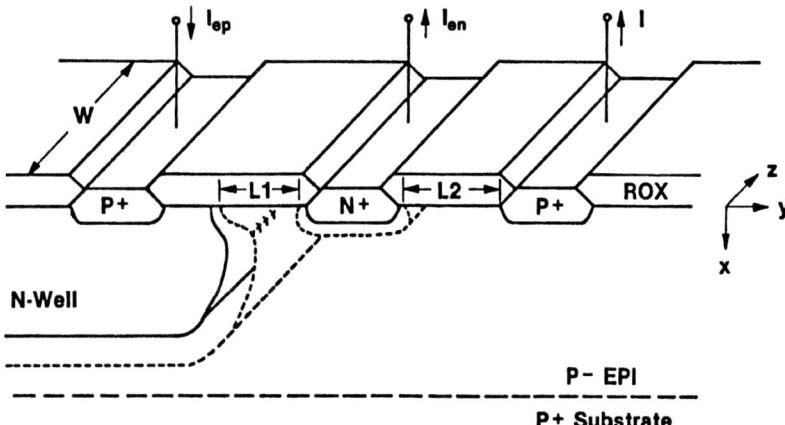

Figure 4-15. Cross section of parasitic PNPN Structure. The ohmic contact to the N-well is omitted for clarity, and depletion region edges are shown by dashed lines. From [Troutman-86]. © 1986 IEEE. Reprinted with permission.

A cross section for the parasitic PNPN structure is shown in Figure 4-15, where the ohmic contact to the N-well is omitted for clarity. A key assumption in the derivation of the transmission line model is that current flow in the z-direction is unimportant. As such, the model describes a two-dimensional potential and current distribution in the x-y plane. This assumption is certainly valid when the width W exceeds the section lengths L1 and L2. When either section length (or both) exceeds W, this approximation provides a worst-case analysis for predicting latchup since it overestimates transfer resistance.

Figure 4-16 shows the corresponding transmission line model for calculating substrate current and potential distributions. Its sections begin at the edge of the depletion region as measured at the surface. The characteristic length and impedance have been derived and discussed in [Troutman-86]. The presence of an $N^+$ diffusion between the N-well and substrate contact interrupts the highly conductive field region and drastically alters the surface potential distribution, and this effect is modeled by including the insertion resistance $R_n$. The line's output resistance $R_2$ depends on the size of the topside substrate contact and can be reduced by using a large area contact. Resistors $R_{w1}$ and $R_{w2}$ account for ohmic drops in the N-well caused by the collected electrons flowing to the N-well contact. Because the collected electrons are constrained to flow within the relatively shallow (compared to substrate thickness) N-well, those resistors usually can be accurately calculated knowing the sheet resistance of the well and the relevant number of squares.

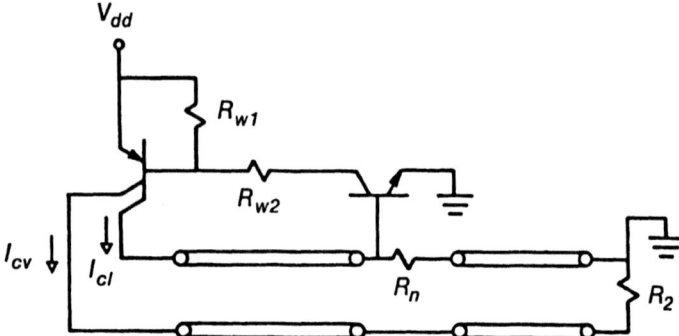

Figure 4-16. Transmission line model for parasitic PNPN Structure. $R_n$ is the insertion resistance of the parasitic $N^+$ emitter, and $R_2$ is the total resistance from the topside $P^+$ diffusion to the $P^+$ substrate ($R_p$ in parallel with $Z_o$).

PNP collector current can excite the line two ways. The majority of holes injected from a $P^+$ diffusion into the N-well flow downward across the base to be collected at the bottom of the N-well. (In many cases all the holes contributing to collector current are collected at the bottom.) This collector current flows through the highly doped substrate and excites the output of the transmission line as it exits a topside substrate contact. Secondly,

## Latchup Models and Analyses

if the $P^+$ diffusion is located close to the N-well/substrate-field junction, a significant number of holes flow laterally through the N-well and are collected at the N-well edge near the surface.[6] This component of collector current excites the input of the transmission line and can also be enhanced by the lateral electric field created in the N-well when a substantial number of electrons are collected by the N-well's side junction. Such a field steers injected holes laterally toward the N-well edge and tends to concentrate hole injection at the emitter edge closest to the collecting junction.

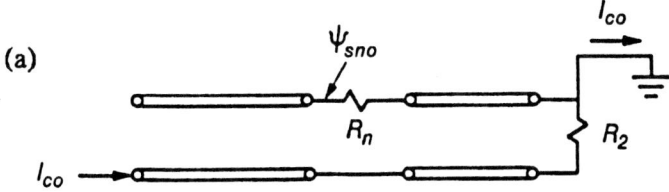

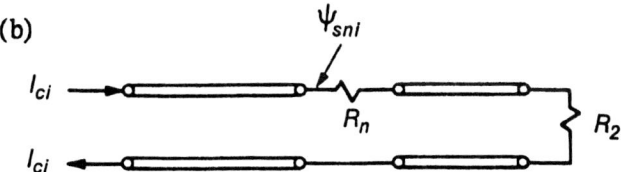

Figure 4-17. Transfer resistance definitions. (a) Output excitation only. Output-excited transfer resistance is defined as $R_{to} = \psi_{sno}/I_{co}$. (b) Input excitation only. Input-excited transfer resistance is defined as $R_n = \psi_{sni}/I_{ci}$.

The two excitations are shown separately in Figure 4-17, where $I_{co}$ is the total collector current measured at the topside substrate contact, and $I_{ci}$ is the component flowing laterally in the N-well and exciting the input. Each excitation produces a surface

---

[6] Takacs, Werner, Harter, and Schwabe have demonstrated that a VNPN has a lateral component when the internal $N^+$ diffusion is close to the P-well edge and that this component can be enhanced by the presence of a field PMOSFET gate electrode [Takacs-84].

potential in the vicinity of a parasitic LNPN emitter, and the general solution is a superposition of the two.

The result of current excitations on the transmission line can also be viewed as two dependent voltage sources on the base of the lateral NPN, as illustrated in Figure 4-18. The equivalent collector resistors are functions of the transmission line parameters. Although expressions for $R_{co}$ and $R_{ci}$ can be obtained directly from the transmission line model, they would be of no value in predicting the switching current. Instead, we want to concentrate on the "transfer" resistances $R_{to}$ and $R_{ti}$ giving rise to $\psi_{sno}$ and $\psi_{sni}$. The first is the change in surface potential at the $N^+$ emitter caused by output current excitation. From the transmission line equations it is easy to show that $\psi_{sno} = R_{to}I_{co}$. The second is the change in surface potential caused by input current excitation. Similarly, $\psi_{sni} = R_{ti}I_{ci}$.

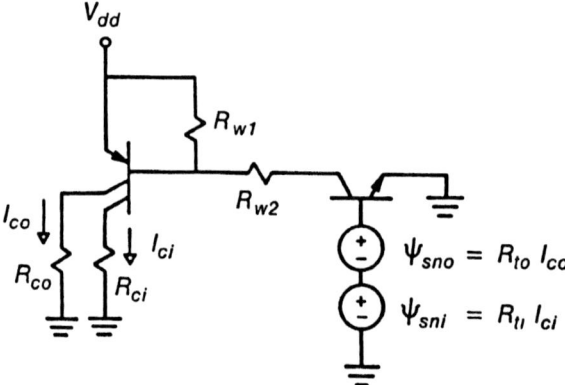

Figure 4-18. Dependent voltage source model of latchup. Each voltage results from a transfer resistance.

The results of the transmission line model can be easily incorporated into the lumped element model by defining an equivalent LNPN bypass resistance that produces the same VPNP collector current. This equivalent bypass resistance is

$$R_{s,eq} = \frac{\psi_{sno} + \psi_{sni}}{I_{co} + I_{ci}} = \frac{R_{to} + \varepsilon R_{ti}}{1 + \varepsilon}, \qquad (4-94)$$

## Latchup Models and Analyses

where $\varepsilon = I_{ci}/I_{co}$, and can be used in the differential latchup criterion for any lumped element calculations such as those performed earlier in this chapter. Depending on where VPNP collector current is actually collected, the equivalent bypass resistance lies somewhere between the transfer resistances $R_{co}$ and $R_{ci}$. These transfer resistances deserve a closer look.

### 4.7.2 Transfer Resistance

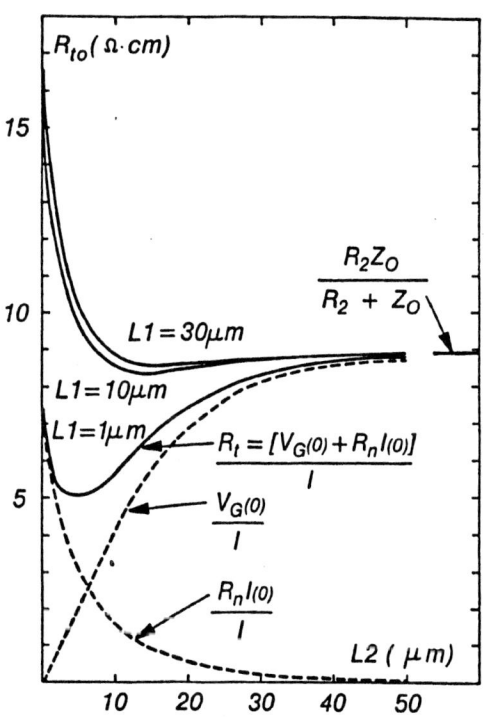

Figure 4-19. Output-excited transfer resistance for high output resistance. The two components are shown for L1 = 1 μm and $R_2$ = 36.6 Ω-cm.

For the case of output excitation Figure 4-19 depicts the transfer resistance dependence on the lengths of transmission line sections on both sides of the $N^+$ diffusion. Plotted on the vertical scale is the surface potential at the edge of the $N^+$ diffusion closest to the N-well divided by the current flowing from the topside substrate contact, which is defined as output transfer resistance $R_{to} = \psi_{sno}/I_{co}$. For the particular output resistance chosen ($R_2 = 36.6$ Ω-cm), transfer resistance first decreases as the substrate contact is moved away from the $N^+$ diffusion, then increases to approach an asymptotic limit.

The two components comprising output transfer resistance are shown by the dashed curves for a 1 μm L1 line. The voltage at the input end of the L2 line increases monotonically with its length and approaches an asymptotic limit when L2 $\gg \gamma$. The current into the L2 line at the input end decreases monotonically, as does the drop across the insertion resistance $R_n$. Consequently, the asymptotic limit for $R_{to}$ at large L2 is the parallel combination of the characteristic impedance $Z_o$ and output resistance $R_2$. It is also clear from this behavior that electron injection from the external $N^+$ diffusion is initially concentrated at the edge adjacent to the N-well and becomes more uniformly distributed over the emitter area as the substrate contact is moved further away.

Output transfer resistance also increases as the length of the L1 line is increased. Looking back toward the N-well from the $N^+$ diffusion, one sees an impedance $Z_{in} = Z_o \coth(L1/\gamma)$, which changes from an open circuit (for L1 = 0) to the characteristic impedance $Z_o$ (for L1 $\gg \gamma$). This impedance, in series with $R_n$, is the input terminating resistance $R_1$ for the L2 line. In the limit of zero L2 (a butted substrate contact), the resistance $Z_{in} + R_n$ is in parallel with $R_2$, and the output transfer resistance, as measured at the node between $R_n$ and $Z_{in}$, can then be written as

$$R_{to} = \frac{R_n R_2}{R_n + R_2 + Z_{in}}. \qquad (4-95)$$

Thus, increasing L1 causes $Z_{in}$ to decrease and the vertical axis intercept of Figure 4-19 to increase.

Latchup protection is gained by minimizing transfer resistance. Output-excited transfer resistance can be minimized by limiting the output terminating resistance $R_2$. Figure 4-20 illustrates the case for $R_2 = 1$ Ω-cm. The worst-case value is now $R_2 Z_o/(R_2 + Z_o)$, which is less than 1 Ω-cm. For a short L1 line it

# Latchup Models and Analyses

is even less. Again the two components of transfer resistance for the L1 = 1 μm case are shown by dashed lines.

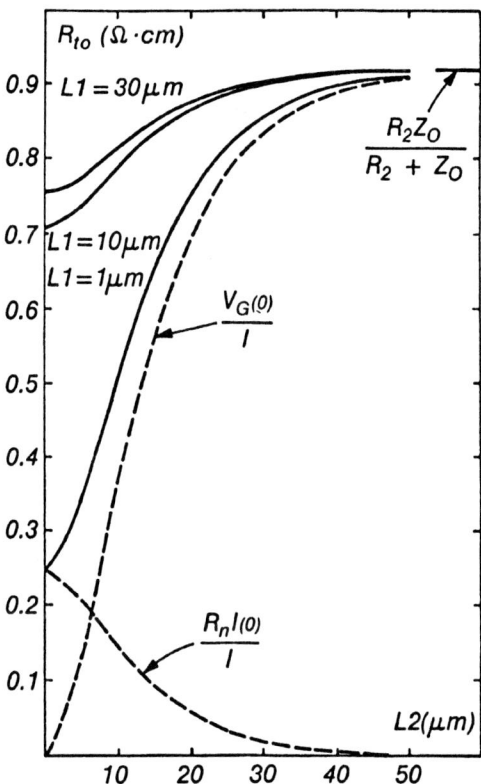

**Figure 4-20.** Output-excited transfer resistance for low output resistance. The two components are shown for L1 = 1 μm and $R_2$ = 1 Ω-cm.

For the case of input excitation Figure 4-21 depicts the transfer resistance dependence on the lengths of the transmission line sections. Once L2 >> γ, the right hand section can be replaced by the characteristic impedance, and there is no further dependence on L2. Lengthening L1 diminishes the current flowing through $R_n$, thus decreasing transfer resistance. Along with small L1, the worst case situation now depends on the output terminating resistance $R_2$. The biggest difference occurs for L2 = 0, where the L1 line is terminated by $R_n + R_2$. For

$L2 \gg \gamma$, it is terminated by $R_n + Z_o$, and there is no dependence on $R_2$. Thus, if $R_2 > Z_o$, the worst case occurs for small L2 (a butted substrate contact) while if $R_2 < Z_o$ the worst case occurs for large L2. For the same VPNP collector current exiting the substrate contact, the surface potential generated increases with the lateral component of well current. Consequently, the trigger current should decrease as the internal emitter is moved closer to the N-well edge. In fact, this has been reported by Lewis in his shallow P-well study [Lewis-84].

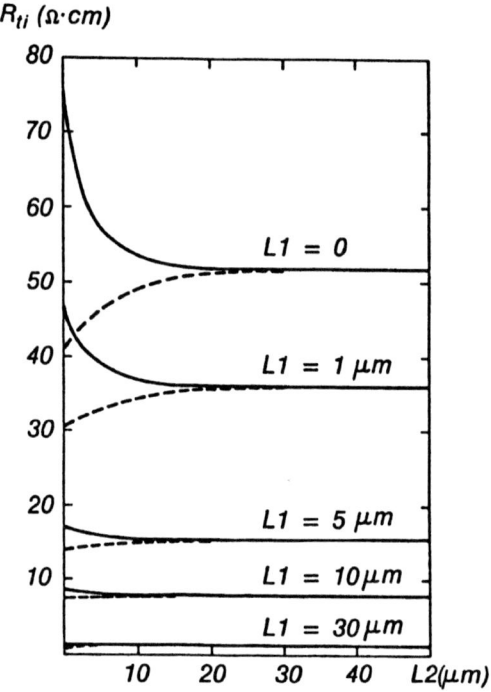

Figure 4-21. Input-excited transfer resistance vs. section lengths. Solid line, $R_2 = 36.6$ Ω-cm; dashed line, $R_2 = 1$ Ω-cm.

Unlike the case of output-excited transfer resistance, simply minimizing $R_2$ does not necessarily reduce $R_{ti}$ to safe values. Consequently, lateral base current in the well should be minimized by other means, such as sufficient spacing of the $P^+$ emitter from the N-well edge or trench isolation at the N-well edge. Alternatively, input-excited transfer resistance should be reduced by using a majority carrier guard in the substrate, which is discussed next.

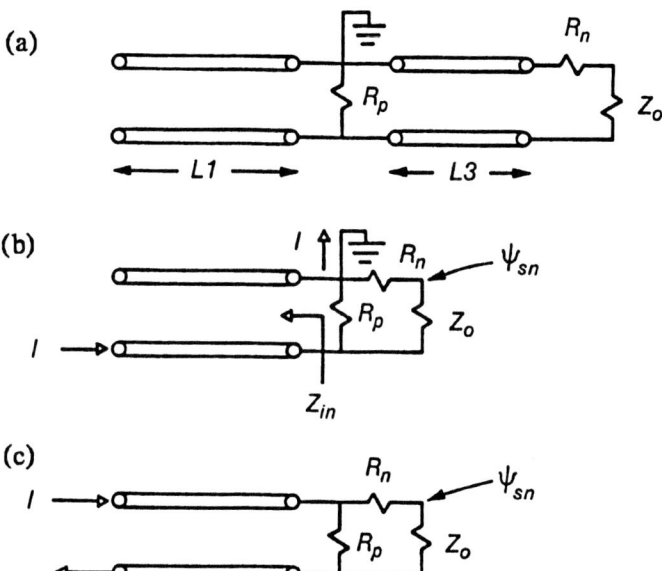

Figure 4-22. Substrate majority carrier guard structure. (a) Transmission line model. (b) Worst-case (L3 = 0) equivalent circuit for output current excitation. (c) Equivalent circuit for input current excitation when L3 = 0.

A second PNPN structural configuration occurs when the topside substrate contact lies between the N-well and the external $N^+$ diffusion, and the corresponding transmission line model is shown in Figure 4-22(a). The substrate is assumed to continue for several characteristic lengths beyond the $N^+$ diffusion so that the $N^+$ insertion resistance is in series with the characteristic impedance $Z_o$. The resistor $R_p$ accounts for the resistance from the topside substrate contact to the highly doped substrate. This transmission line model is similar to the case discussed above, but, for output current excitation, surface potential is now larger on the $N^+$ diffusion edge away from the N-well. Again the worst-case output-excited transfer resistance occurs for a butted contact, or when the section length L3 = 0. For this situation the lumped element model of Figure 4-22(b) is applicable. Output transfer resistance is now expressed as

$$R_{to} = \frac{R_n R_2}{R_n + R_2 + Z_o}, \qquad (4-96)$$

where now $R_2 = Z_{in}R_p/(Z_{in} + R_p)$. Once again the design goal is to make $R_2$ as small as practical, which is accomplished by minimizing $R_p$.

Figure 4-22(c) shows the equivalent circuit when L3 = 0 for input current excitation. The direction of current flow through the insertion resistance $R_n$ now causes a negative substrate potential on the side of the $N^+$ diffusion away from the N-well so that input excitation does not cause a problem when the topside substrate contact is between the N-well and an external $N^+$ diffusion. (For non-zero L3, the substrate potential on both sides is negative.)

Placing a substrate contact between the N-well and an external emitter (in effect adding a majority carrier guard structure) does not, in itself, prevent latchup in all cases [Ochoa-81]. Nonetheless, there are significant benefits. Rung and Momose reported that in a PNPN structure with both external and internal $N^+/P^+$ pairs reversed, so that the ohmic contacts to the P-well and N-type substrate were the closest diffusions, holding current was significantly increased, even when the P-well was left floating [Rung-83]. Systematic studies comparing regular and reversed diffusions have also been reported in [Troutman-84] and [Hu-84ab].

## 4.8 Dynamic Latchup Effects

Previous sections have ignored the time dependence of latchup and concentrated on developing latchup criteria and discussing the lumped element model for the PNPN device and its extensions. There are several important time-dependent considerations regarding latchup, and in this section we investigate the time-dependent, or dynamic, nature of latchup.

### 4.8.1 Sources of Time Dependence

There are two effects limiting the latchup response time of a PNPN device to a time-varying excitation. First is the base transit time of each transistor, which limits how quickly a transistor's collector can build up after a step excitation at the base/emitter junction. Second are the various time constants inherent to the structure that limit how quickly an excitation can actually be applied to the emitter/base junction.

The minimum time for regeneration to occur, assuming an instantaneous base/emitter voltage at both transistors, is given by the sum of the base transit times $\tau_n + \tau_p$. When minority carrier diffusion length is large compared to basewidth W,

$$\tau_{n,p} = \frac{W_{n,p}^2}{2D_{n,p}} \qquad (4-97)$$

at low injection levels, and

$$\tau_{n,p} = \frac{W_{n,p}^2}{4D_{n,p}} \qquad (4-98)$$

at high injection levels when base transit is by diffusion only. Figure 4-23 compares the sum of the two base transit times with minimum turn-on time [Fang-84]. For lateral base widths greater than 18 $\mu$m the data agrees better with the curves using a lifetime for the LNPN including high level injection. For shorter base widths the data appears to agree better with the curve for low level injection only. The offered explanation is that at shorter basewidths, base current flows close to the surface, where doping density is relatively high ($2E16/cm^3$), while for the longer basewidths, base current spreads more into the substrate where doping density is much lower. Alternatively, because of this current spreading, the effective base width could easily be greater than the distance from the $N^+$ emitter to the N-well, which would shift the low level injection curve closer to the data.

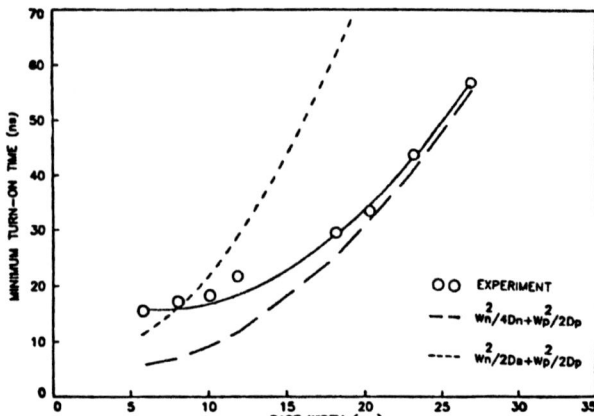

Figure 4-23. Comparison of turn-on time for high and low level injection. The VPNP is pulsed and minimum turn-on time is measured vs. LNPN base width. From [Fang-84]. © 1984 IEEE. Reprinted with permission.

Because the lateral device usually has the larger base width for the two parasitic bipolars, it limits the transient response. However, field-aided base transport, which is present for some types of triggering, can reduce the transit times below those predicted by the above equations, and field-aided base transport can be an important effect in lateral bipolar transistors [Estreich-80].

Even if the transistors themselves are assumed to respond immediately to any excitation (zero base transit time), the response of the PNPN structure to fast triggering can still be finite. This has been illustrated by Troutman and Zappe, who investigated the response of a PNPN device to a fast, linear ramp in the power supply [Troutman-83]. The transistor models used in this study were deliberately oversimplified to emphasize the fundamental time dependencies in latchup. Not only was transistor current gain taken to be a constant, but the diodes were approximated with a piecewise linear characteristic, which permitted analytic solutions. A transistor was "off" if its base/emitter bias was below some critical voltage $V_{b,on}$ and "on" if it equaled or exceeded $V_{b,on}$. A linear voltage ramp was chosen to keep the algebra reasonably simple. This ramp produced a time-dependent displacement current through the well/substrate

## Latchup Models and Analyses

junction capacitor C (assumed constant for simplicity)[7] and through the bypass resistors $R_1$ and $R_2$.

This displacement current flowing through the bypass resistors, if large enough, can trigger latchup. Until the first transistor (the one having the larger bypass resistor) turns on, the relevant time constant for the buildup of total current into the PNPN device is $\tau_1 = (R_x + R_1 + R_2)C$, where $R_x$ is the total resistance in series with the PNPN structure, $R_1$ is the bypass resistor for the first transistor to turn on, and $R_2$ is the bypass resistor for the second. When the current into the PNPN device reaches a value $I_{1,on} = V_{b,on}/R_1$ the first transistor turns on, and the appropriate time constant becomes $\tau_2 = (R_x + R_2)C/(1 - \alpha_1)$. In effect the junction capacitance C is now multiplied by feedback supplied by the "on" transistor, and when $\alpha_1$ is close to one, this effect can result in a significantly slower time scale. When the current reaches a value $I_{2,on} = V_{b,on}/R_2$, the second transistor turns on. The response function now includes a rising exponential (assuming $\alpha_1 + \alpha_2 > 1$), and the appropriate time constant is $|\tau_3| = R_x C/(\alpha_1 + \alpha_2 - 1)$. Current continues to rise until the switching point is reached at a value[8] of $I_s = (\alpha_1 I_{1,on} + \alpha_2 I_{2,on})/(\alpha_1 + \alpha_2 - 1)$, at which point the reverse bias on the well/substrate junction begins to decrease.

If the excitation is terminated before the switching point is reached, even after both transistors are turned on, the current decays, and the PNPN device returns to the blocking state. This dynamic recovery behavior is shown in Figure 4-24(a). On the other hand, once the switching point is reached, the device moves into the latched state (to a point determined by the load line $1/R_x$ and the final value of the voltage ramp), even when the dynamic

---

[7] A numerical analysis using a voltage-dependent well/substrate junction capacitance has extended the range of ramp rates for which latchup does not occur [Fu-85]. This result follows from the fact that junction capacitance decreases as reverse bias builds up, reducing the displacement current accompanying the ramp.

[8] The more rigorous definition of switching current provided by the differential latchup criterion discussed in section 4.4 can be used to ascertain the appropriate $V_{b,on}$ values.

excitation is terminated. This dynamically triggered latchup behavior is pictured in Figure 4-24(b).

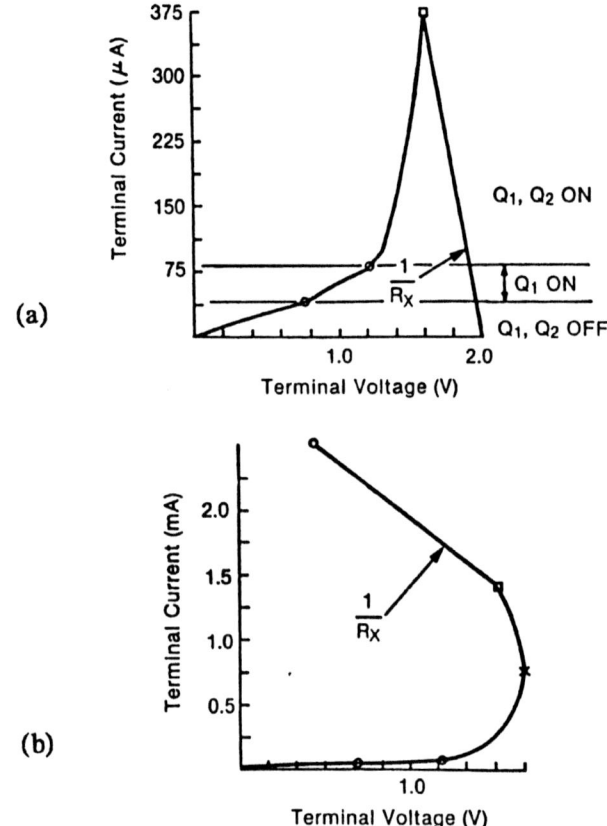

Figure 4-24. Simulated PNPN response to ramped power supply. Transistors are simulated by piecewise linear, abrupt turn-on model. (a) Dynamic recovery. (b) Latchup. From [Troutman-83]. © 1983 IEEE. Reprinted with permission.

The same work also reports on numerical simulations of the dynamic latchup behavior of the PNPN using more realistic transistor models. Although there are no abrupt changes identifying transistor turn-on in the numerically simulated waveform as there are for the analytical model, results are essentially identical. In particular, the numerical simulations also reveal dynamic recovery effects. Finally, both this work and subsequent work by the same authors [Troutman-84] have experimentally verified dynamic recovery using a ramped voltage for the power supply.

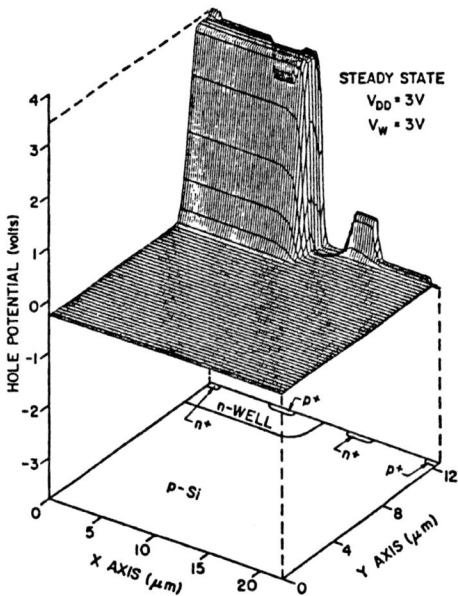

Figure 4-25.  Blocking state potential distribution.  From [Hu-84a]. © 1984 IEEE. Reprinted with permission.

The significance of dynamic recovery is its clear demonstration that while turning on both transistors in a PNPN device is a necessary condition for latchup, it is not a sufficient condition. More fundamental to latchup is whether the switching point has been reached. Whether the latchup dynamics are limited by base transit time or by RC terms, the critical response time is the time required for the PNPN terminal current to reach the switching current.  If the excitation is terminated before the switching current is reached, the PNPN device returns to a steady-state operating point in the blocking state. Otherwise, latchup results. The total time to reach the latched state also includes the time to discharge the well/substrate junction and forward bias this junction sufficiently to reach the steady-state point in the latched state.

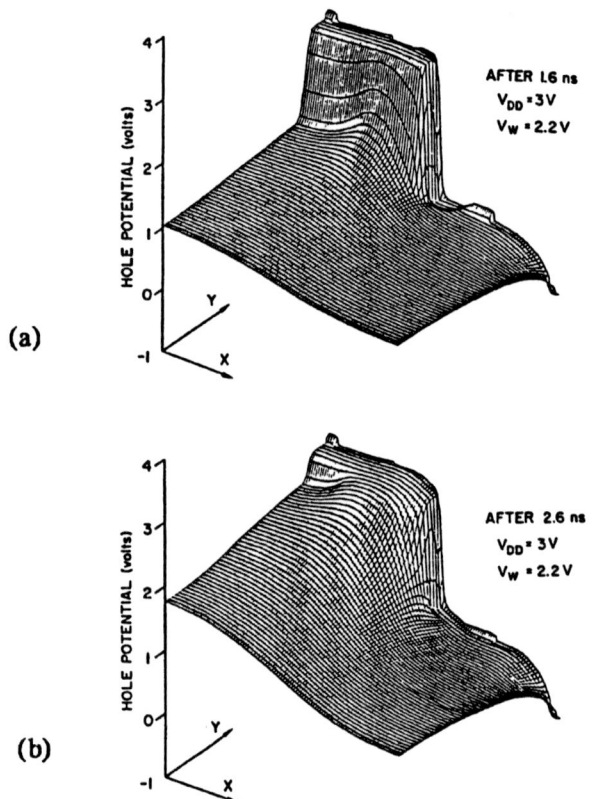

**Figure 4-26.** Transient potential distribution. (a) 1.6 nS and (b) 2.6 nS after pulsing VPNP base. The emitter/base forward bias is 0.8 volt. From [Hu-84a]. © 1984 IEEE. Reprinted with permission.

Latchup transient behavior has also been investigated using two-dimensional simulations.[9] These simulations reveal the physical changes inside the PNPN device as it switches from the blocking state. Hu has studied the PNPN response to a pulse excitation on the VPNP base [Hu-84a]. For reference, Figure 4-25 illustrates the steady-state potential distribution in the blocking state. Figure 4-26(a) shows the instantaneous

---

[9] Two dimensional numerical simulations have also been used to investigate static excitation. See [Wieder-81], [Taur-84], and [Pinto-85] for examples.

potential distribution 1.6 nS after the N-well potential has been lowered from 3 to 2.2 V at t = 0. There is a large hole current flowing in the substrate, and the LNPN has been heavily forward biased. Nonetheless, the total current into the PNPN device at this time is less than the switching current, and if the N-well is now returned to 3 V, the PNPN device recovers to the blocking state. Figure 4-26(b) shows the potential distribution after 2.6 nS with the forward bias still applied. The large hole current has driven a large portion of the N-well bottom junction into saturation. The external $N^+$ diffusion is so heavily biased it is barely visible, and it prevents the side of the N-well from also going into saturation. At this point the VPNP is in saturation, but the LNPN is still operating in the forward active region. Total device current exceeds the switching current, and if the N-well is returned to 3 V, the device does not recover. Instead, the current continues to rise until limited by the appropriate load line.

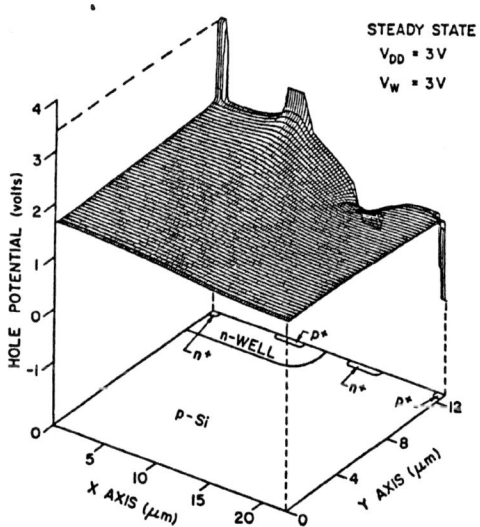

Figure 4-27. Latched state potential distribution. Terminal current is nearly 1 A/μm of width for the applied voltages. From [Hu-84a]. © 1984 IEEE. Reprinted with permission.

Assuming the device could handle the simulated current level for zero external load resistance (approximately 1 A/cm), the resulting potential distribution would be that shown in Figure 4-27. The N-well is no longer distinguishable. Both substrate and well are flooded with holes and electrons ($\simeq 1E19/cm^3$), and the applied voltage is dropped primarily across the resulting space charge.

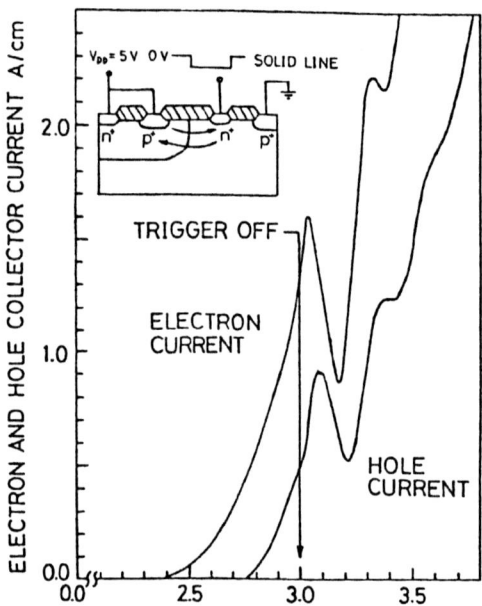

Figure 4-28. Electron and hole currents vs. time in nS. From [Odanaka-85]. © 1985 IEEE. Reprinted with permission.

As for undershoot on the other parasitic transistor, Odanaka, Wakabayashi, and Ohzone have studied the PNPN response to a pulse excitation on the LNPN emitter [Odanaka-85]. Figure 4-28 shows the electron and hole currents at the well junction as a function of time. The electron current becomes visible on this scale 2.4 nS after the pulse is applied and rises sharply. Hole current becomes visible after 2.75 nS, which is the estimated sum of both base transit times for low level injection. The electrostatic potential at 2.9 nS, shown in Figure 4-29(a), reveals that hole current does not yet modulate substrate potential, so that electron transport is still diffusion-limited at 2.9 nS. The electron density around the LNPN at this time, shown in Figure 4-29(b), exceeds substrate doping by an order of magnitude, indicating the onset of high-level injection effects. Electron current density has doubled over that at 2.75 nS, so that high level injection effects are just beginning at 2.75 nS.

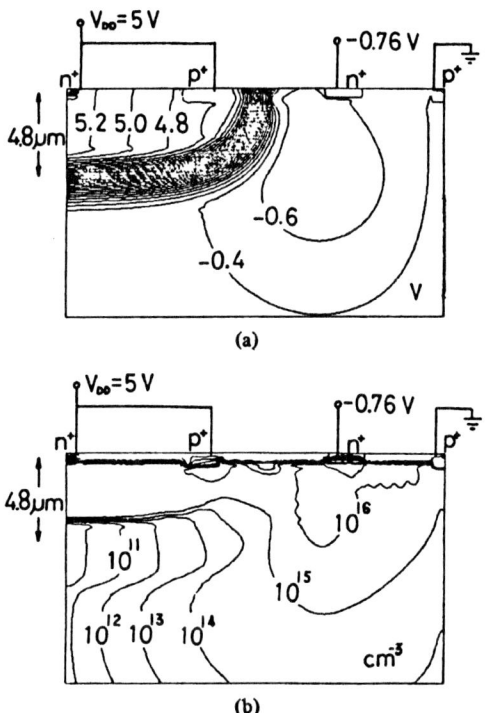

Figure 4-29. Instantaneous behavior 2.9 nS after pulsing LPNP base. (a) Electrostatic potential. (b) Electron density. From [Odanaka-85]. © 1985 IEEE. Reprinted with permission.

Figure 4-30(a), (b), and (c) show potential distribution, electron current density, and hole current density at t = 3.4 nS. Hole current injected from the $P^+$ emitter flows laterally to the N-well edge and now causes an appreciable lateral electric field (>1 kV/cm), which enhances electron collection at the N-well edge facing the external $N^+$ emitter. Initially, injected electrons flowed more deeply into the substrate, and many were collected on the bottom side of the well junction. Now, after latchup, they follow a more direct path to the side of the N-well. Electron and hole currents will continue to increase until limited by an external load line.

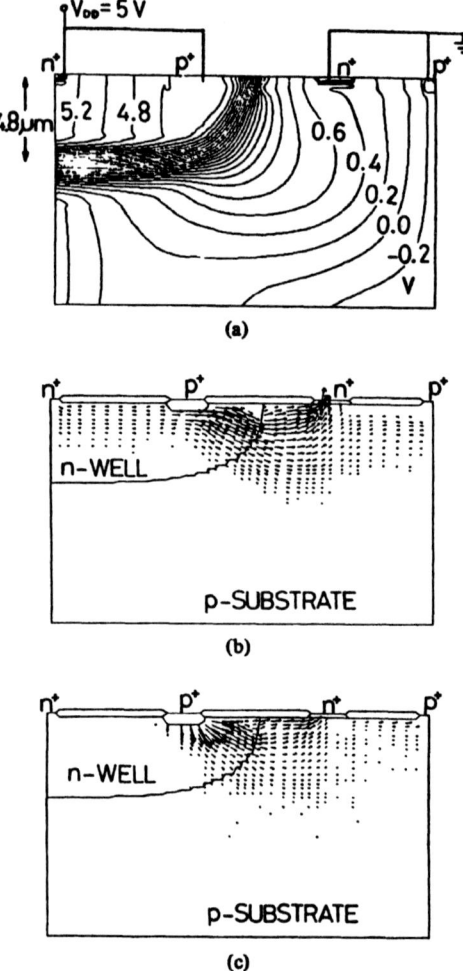

Figure 4-30. Instantaneous behavior 3.4 nS after pulsing LNPN base. Pulse was turned off at 3.0 nS (a) Electrostatic potential. (b) Electron current density. (c) Hole current density. From [Odanaka-85]. © 1985 IEEE. Reprinted with permission.

A reciprocal effect has been demonstrated for P-well CMOS [Pinto-85]. The LPNP collector current flowing through the well creates a lateral electric field that concentrates VNPN emitter injection on the side closest to the P-well edge and acts to steer injected electrons toward the P-well's side junction. This effect

Latchup Models and Analyses 115

emphasizes the fact that when stimulated by the lateral parasitic bipolar, the vertical bipolar can behave differently than when externally biasing its base/emitter. In some cases the choice of transistor parameter values must account for such synergetic effects when estimating the switching current.

Modeling the transient behavior of latchup has graphically demonstrated the changes in potential and charge distributions inside the PNPN structure. It has also revealed why transistor action can drastically change when the two parasitic devices interact. In spite of these complications, the differential latchup criterion introduced in section 4.3.3 is relevant to transient triggering. Its application now requires modeling the transient response of the PNPN structure to determine whether total PNPN current reaches the switching level. Demonstrating that it does not under realistic conditions insures a latchup-free design.

## 4.9 Modeling and Analysis Review

Latchup analyses are all based on some understanding of when latchup occurs, i.e., what conditions must hold before a parasitic PNPN device switches from the blocking state to the latched state and how those conditions can be described mathematically. Descriptions of latchup criteria given in the technical literature have been based on various versions of the loop gain inherent to the PNPN device. Surprisingly however, there has been little attempt to analytically test the proposed criteria.

This chapter has introduced the differential latchup criterion (DLC), which is a precise and rigorous statement of when latchup does and does not occur, and has demonstrated its validity with an analytical example. This criterion is a compact statement of the loop gain requirement that includes the effects of bypass resistance as well as transistor current gain. It has no fitting parameters and can be applied to any parasitic PNPN structure. In its most general form, the differential latchup criterion also accounts for series base and emitter resistance, high level injection effects in the parasitic bipolars, and distributed substrate resistance.

Application of the differential latchup criterion is facilitated by the triode approximation, which replaces the general tetrode with a PNPN triode formed by removing the larger bypass resistor. This approximation also leads to a simple and accurate equation for switching current. It is always a valid approximation as long as the trigger circuit is properly included in the PNPN configuration. When one bypass resistor is more than twice as large as the other, it predicts results within a few per cent of those calculated numerically.

While PNPN loop gain is central to understanding latchup, other considerations are also important. Excitation of the first bipolar must endure long enough for the complementary parasitic bipolar to turn on and develop the necessary loop gain. This effect has been demonstrated various ways by several authors. Dynamically, both RC time constants and base transit time can have a measurable effect on the transient response of the PNPN device.

This chapter has also examined the holding point and derived new equations for holding current and voltage. These equations are the first to include reverse transistor behavior, an important consideration since at least one parasitic bipolar is heavily saturated at the holding point and a consideration that has been largely overlooked in the technical literature.

Latchup cannot occur as long as all parasitic PNPN devices operate solely in SAFE space, i.e. as long as $(\alpha^*_{fns} + \alpha^*_{fps}) < 1$. Understanding this inequality and how to apply it is the key to avoiding latchup.

# Chapter 5

## LATCHUP CHARACTERIZATION

In the previous chapter we learned the SAFE space switching boundary is rigorously defined by the differential latchup criterion. It is time to explore the experimental techniques for measuring this boundary.

Although latchup has been studied for many years, there is no universal technique for characterizing its behavior. In fact, there is not even a standard terminology for the experimental observations. Different investigators have used the terms trigger current, critical current, switching current, and holding current to mean different things. Latchup behavior depends on the structure exhibiting latchup, the method of triggering latchup, and, of course, the external conditions of all the terminals under investigation. Useful and accurate comparisons of experimental results are possible only if the characterization conditions are carefully specified.

This chapter reviews characterization techniques, discusses their results, and evaluates their utility. Following a brief discussion of the two measurement tools used for virtually all static latchup characterization, it systematically elaborates the terminal connections and measurement conditions for static characterization. Characterizing the switching and holding points is given special attention since they are especially important for measuring latchup hardness. In addition, new data shows switching point measurements over a wide range of bypass resistances to be in excellent agreement with the differential latchup criterion discussed in Chapter 4, which we have already noted has no fitting parameters.

The various dynamic characterization techniques are also reviewed. Probably the most practical goal here is understanding the triggering conditions of transient waveforms, especially the kinds of waveforms seen by off-chip receiver and driver circuits. The chapter ends with a discussion of latchup temperature dependence and non-electrical latchup characterization.

## 5.1 Measuring Instruments

Care must be exercised in making latchup measurements to insure that applying voltages and/or currents to the terminals of a PNPN structure produces the desired excitation. The bulk of this chapter is concerned with static triggering; the exciting voltages and/or currents are applied statically, and the consequent behavior is a result of the terminal conditions themselves, not their time rate of change. In static triggering, a voltage source should be changed slowly enough that displacement currents do not trigger latchup (a form of transient triggering), and in the case of a programmed voltage source, the voltage change on the staircase output should be limited to avoid significant displacement currents. Likewise, current source changes must avoid inductive voltages large enough to induce triggering.

### *5.1.1 Curve Tracer*

Until relatively recently latchup was characterized primarily using a curve tracer, and curve tracers are still widely used. Current/voltage characteristics showing latchup in a parasitic PNPN device can be traced relatively quickly, just as for a bipolar or field effect transistor, and the curve tracer's current limiting capabilities can be used to prevent permanent damage when the PNPN structure is switched into the low impedance state. Because a trace is made both when the voltage is swept from zero to its maximum value and when it is decreased again to zero, the entire blocking and latched states are exhibited. In many cases the curve tracer's measurement speed is high enough to measure portions of the negative differential resistance region as well, which aids identification of the zero differential resistance points.

# Latchup Characterization

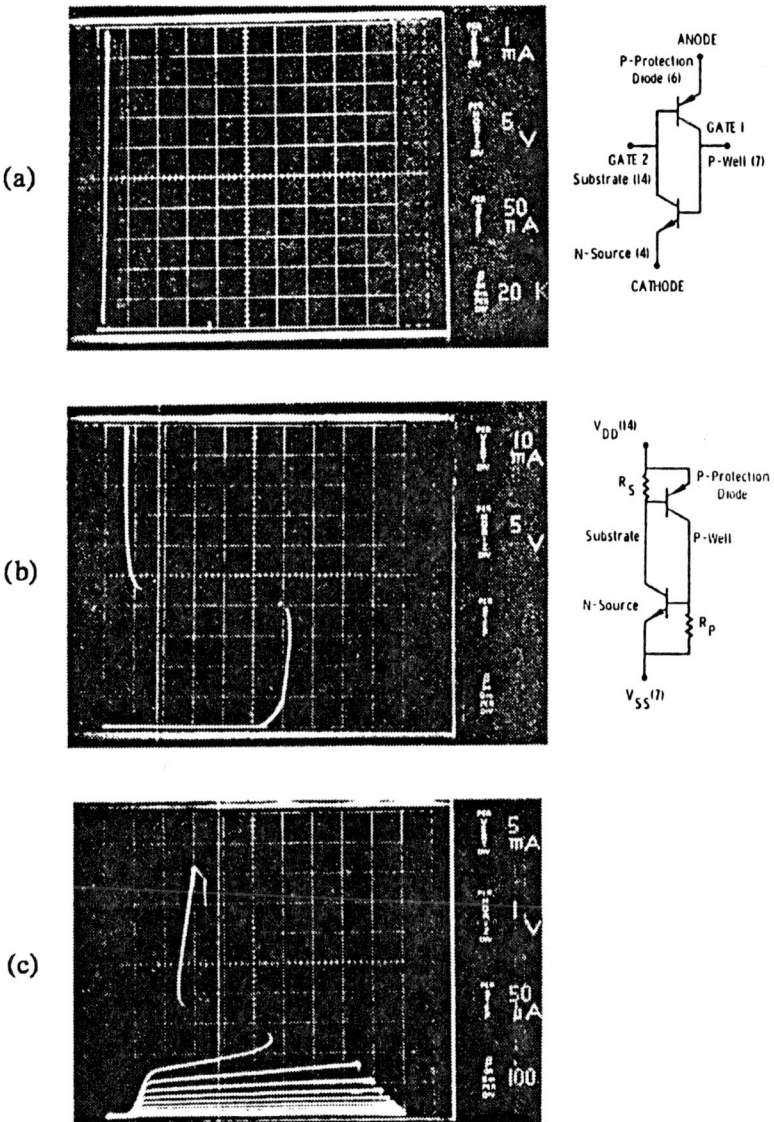

Figure 5-1. Curve tracer characterization of latchup. (a) Supply overvoltage triggering for the diode configuration. (b) Supply overvoltage triggering for the tetrode configuration. (c) Gate triggering using the P-well (a triode configuration). From [Gregory-73]. © 1973 IEEE. Reprinted with permission.

Early parasitic PNPN characterization using a curve tracer was based on similar characterization of SCR devices. In one technique the curve tracer's voltage sweep was used to implement supply overvoltage triggering with no SCR gate current applied and was a direct test of high voltage triggering. Gregory and Shafer applied this supply overvoltage technique to both the PNPN diode configuration, with the results shown in Figure 5-1(a), and to the tetrode configuration, with the results shown in Figure 5-1(b) [Gregory-73]. The sum of the small-signal common base gains for the two parasitic transistors is less than unity since the diode configuration exhibits a blocking state. For this sample latchup is triggered by avalanche current.

In a second technique the curve tracer's base drive was used to provide SCR gate current at a fixed anode-to-cathode voltage. In the same work cited above the base drive was applied directly to either the substrate or P-well terminal, since these are the "SCR gates" of the parasitic PNPN structure, and typical results are shown in Figure 5-1(c). This display is essentially the family of VNPN $I_c$ vs. $V_{ce}$ characteristics with the LPNP emitter shorted to the substrate, which acts as the collector for the VNPN. Triggering here is a combination of gate current drive and high collector voltage, and applying this combination switches the sample to the latched state. Note that for gated triggering the parasitic PNPN structure measured has a triode configuration, which is not always a valid representation of the tetrode.

Curve tracer characterization does not always yield SCR-like results for the CMOS parasitic PNPN structure, however. An example of SCR current/voltage characteristics is shown in Figure 5-2(a) as a comparison to the same characteristics for a parasitic PNPN structure shown in Figure 5-2(b) [Dressendorfer-81]. Both the blocking and latched states are visible, and the switching voltages are approximately 8 and 22 volts for the SCR and CMOS parasitic PNPN, respectively. The corresponding holding voltages are approximately 0.8 and 4.5 volts. Recall that in the latched state the total voltage drop for the SCR's equivalent circuit is the forward bias on the base/emitter junction of one transistor ($\simeq 0.5$ V) added to the collector/emitter saturation voltage of the second transistor ($\simeq 0.3$ V). In the above case this equivalent circuit is valid for the

## Latchup Characterization

SCR, but it clearly fails to explain parasitic PNPN characteristics such as those in Figure 5-2(b). In addition, the impedance for the latched state of the parasitic PNPN exceeds that for the SCR and increases at higher current, with its total current saturating at approximately 320 mA. Dressendorfer and Ochoa have attributed these differences to the additional lumped resistors indicated as $R_{s2}$ and $R_{w2}$ in Figure 4-3 on page 47. [Dressendorfer-81].

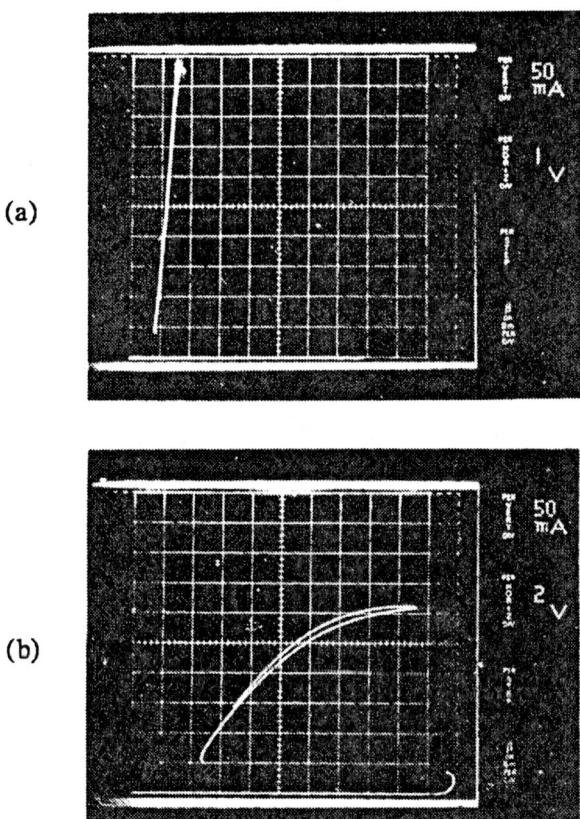

Figure 5-2. Supply overvoltage triggered characteristics. (a) Classical SCR latchup. (b) Parasitic PNPN latchup in a CD 4007 inverter. From [Dressendorfer-81]. © 1981 IEEE. Reprinted with permission.

# Latchup in CMOS

(a)

(b)

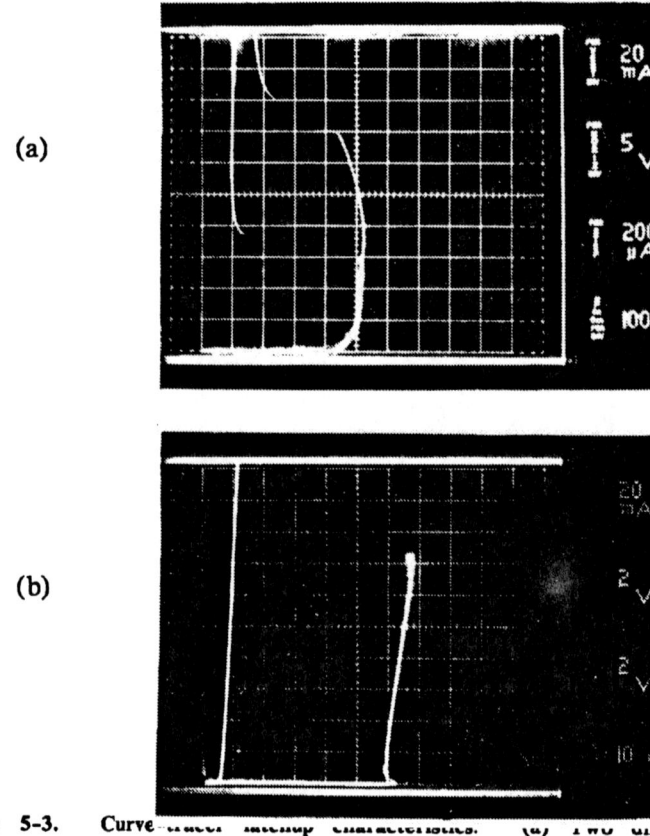

Figure 5-3. Curve tracer latchup characteristics. (a) Two different parasitic PNPN structures. (b) Epitaxial P-well CMOS latchup for floating substrate triode structure (left trace); latchup does not occur for tetrode structure (right trace). From [Ochoa-79]. © 1979 IEEE. Reprinted with permission.

Structural comparisons of various PNPN devices have been made using holding currents read directly from the oscilloscope, and curve tracer plots for a second example of P-well latchup triggered by overvoltage reveal some interesting points [Ochoa-79]. Figure 5-3(a) illustrates the effect of different current paths on the characteristics of a bulk (non-epi) sample. Precisely identifying holding current (as defined by $dV/dI = 0$) on these curves is more difficult than on those in Figure 5-2, but holding current for one is clearly lower than that for the other.

Latchup Characterization

The lower holding current corresponds to latchup between an input line (connected to an external $P^+$ diffusion in the protection circuit, which forms an LPNP emitter) and $V_{ss}$ (connected to the N-well and the lowest potential on the chip), while the higher holding current corresponds to latchup between $V_{ss}$ and the parallel combination of an input line and $V_{dd}$. The parallel combination of input and power supply lines reduces the effective bypass resistance for the two LPNP devices, which together with the VNPN form the latching circuit. With $V_{dd}$ connected to the most positive potential (the usual configuration), latchup does not occur for the epi-CMOS sample shown in Figure 5-3(b), even though current rises sharply because of avalanche breakdown at approximately 12 volts. If $V_{dd}$ is left open, however, increasing $V_{in}$ (now the effective anode voltage) causes latchup. This last test circuit is effectively a triode configuration with the N-type substrate floating.

The second measurement technique (using a curve tracer's base drive to trigger latchup) has also been used to evaluate various latchup fixes. In [Sakai-81] VNPN characteristics measured with an external $P^+$ diffusion shorted to the substrate are compared for non-epi, intrinsically gettered non-epi, and epi-CMOS at four different spacings from the external $P^+$ diffusion to the P-well. Only the epi-CMOS samples were free from latchup. In [Sangiorgi-84] similar characteristics were made for a VPNP with an external $N^+$ diffusion shorted to the substrate and used to compare structures with and without Schottky barrier PMOS source/drains. In this work trigger current was defined as the minimum base (N-well) current which causes latchup for $V_{ce} < 5$ V.

This same characterization technique has also been used on structures more complicated than a parasitic PNPN. In [Dooley-84] the base current step generator was used to sink current from the output of a logic inverter. For such a measurement there are really three bipolar transistors in the circuit - the usual two, plus the one on the output node being used to trigger latchup, a VNPN in this case. Consequently, two latchup paths are possible, and one can trigger the other. Since measured trigger current (output emitter current) and threshold current (the nomenclature in this reference for switching current) differed little when the inverter input was changed from $V_{dd}$ to ground, the measured currents in the referenced work probably represent latchup of the usual two bipolars.

In [Huang-82] three different buffer circuits were compared for latchup hardness. When the buffer input signal $V_{in}$ =5 V, trigger current is forced into the output node, and when $V_{in}$ =0 V, it is forced out. Trigger current was lower for $V_{in}$ =5 V than for 0 V, but for both cases it exceeded holding current for most buffer circuits measured. Curves for these buffers exhibit multiple switching points, indicating that one latchup path can excite another.

### 5.1.2 Parameter Analyzer

Recent characterization work has taken advantage of the flexibility inherent to the HP 4145A parameter analyzer in order to elucidate triggering mechanisms, the exact sequence of events prior to latchup, and current/voltage behavior of the blocking and latched states for the actual operating configuration. This instrument has four system measurement units (SMU) that can simultaneously source current and measure voltage at a terminal (or alternatively, source voltage and measure current). It also has a convenient provision for programming the tests to be performed and displaying the results either graphically or in a printed table. Since the digitized data can be stored in memory or on a disk, it is easily re-displayed using different axes or combined with other data to help identify the physical mechanism behind a particular I/V curve. There are some possible problems, however. See Appendix B for a discussion.

Plots resulting from the emitter current excitations are shown in Figure 5-4. These curves are the blocking state characteristics of the currents into the four external terminals, and are plotted versus the voltage that builds up on the excited emitter. Until the second transistor turns on, such plots simply show base, emitter, and collector current vs. emitter voltage for the externally excited transistor. One can immediately see, for instance, the VPNP common emitter current gain is approximately 75 and that for the LNPN current gain is approximately 3. After the second transistor turns on, device current rises steeply to the switching point, but negative differential resistance is not observable with

# Latchup Characterization

the logarithmic current sweeps used in Figure 5-4. Section 5.4 shows how the switching point can be fully displayed by measuring a portion of the negative differential resistance region in its vicinity.

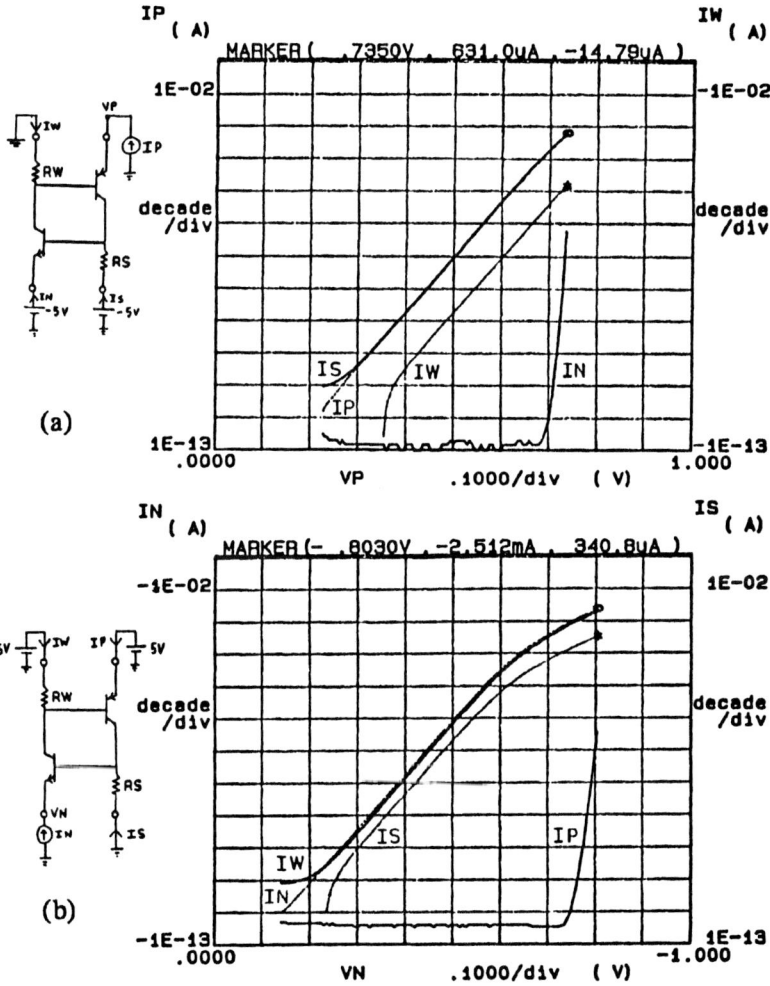

Figure 5-4. Blocking state characterization. Current source in emitter of (a) PNP and (b) NPN. The first transistor is forced on using a current source in its emitter. The second transistor is eventually turned on by current through its bypass resistor. From [Leventhal-84]. Reprinted with permission.

## 5.2 Two Terminal Characterization

In two terminal characterization only two terminals are presented to the measuring apparatus, which could be a curve tracer or a parameter analyzer. Possible measurement configurations include the PNPN diode, the triode with one pair of terminals shorted together (the emitter and bypass resistor of one transistor), or the tetrode with two pairs shorted together. Two terminal triggering is accomplished by raising the anode voltage (or current) source until latchup occurs, and such triggering is then said to be ungated or self-gated and caused by supply overvoltage (or overcurrent) stress. Two terminal characterization is used nearly universally for latched state measurements, regardless of whether triggering was gated or ungated.

### 5.2.1 Supply Overvoltage Stress

As the first example of static latchup triggering, consider the frequently used overvoltage stress on the power supply. Overvoltage on the power supply is commonly used to test for limits or weak points in a design, such as the maximum allowable field implant or the minimum spacing that can be tolerated between a source/drain diffusion and the N-well (or P-well) edge. Voltages somewhat in excess of those to be encountered under actual operating conditions can be used to buffer a design or test the interaction of processing parameters.

Figure 5-5(a) shows the general terminal conditions for a four terminal PNPN structure, with a Thevenin equivalent being used to simulate the power supply circuitry. The external well and substrate resistors $R_{wx}$ and $R_{sx}$ can be used to simulate various spacings of the well or substrate contacts from the nearby emitter diffusion, as well as various values of contact resistance. Alternatively, the intrinsic structure can be tested by setting $R_{wx} = R_{sx} = 0$. Latchup measurements on the inverter shown in Figure 5-5(b) require the additional specification of its input node, which is usually connected to either ground or the power supply voltage.

Latchup Characterization                                                127

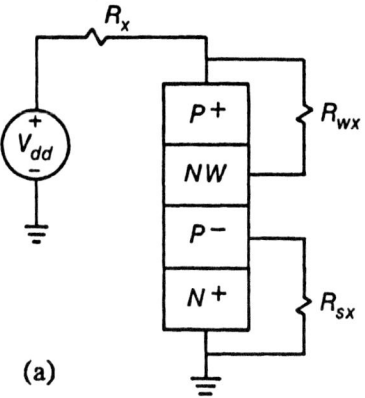

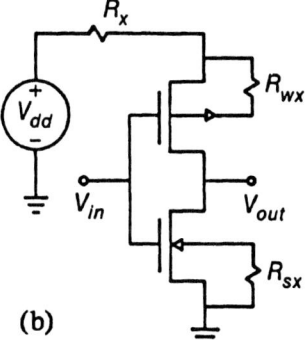

Figure 5-5. Supply overvoltage stress condition. (a) PNPN structure. (b) CMOS inverter. $R_{wx}$ is the external well resistance, $R_{sx}$ is the external substrate resistance, and $R_x$ is the external series PNPN resistance.

Measurement proceeds by slowly raising $V_{dd}$, to avoid displacement currents that could trigger latchup, and by recording terminal voltage V and terminal current I for the latchup structure. If the spacings from either $N^+$ or $P^+$ diffusions to the well edge are large enough to preclude punchthrough, the ultimate limit on $V_{dd}$ for the PNPN structure will be set by well/substrate junction breakdown. (See, for example, [Estreich-80].) Breakdown of this junction provides a current path across the collector/base junction of the parasitic bipolars.

Because of the relatively low surface doping in the field region both inside and outside the N-well (between 1E16 and $1E17/cm^3$), N-well junction breakdown can be quite high, and it is certainly possible to place diffusions close enough to the N-well edge so that punchthrough imposes the limit on $V_{dd}$. Punchthrough could easily impose a testing limit in CMOS designs prizing the minimum possible $N^+/P^+$ spacing.

A third possible limit to the $N^+/P^+$ spacing is field FET triggering [Takacs-84]. Field inversion then provides a current path, which results in latchup for sufficiently large channel current. Because of short channel threshold reduction, this is a serious design consideration in high density CMOS.

If an overvoltage stress is performed on an active inverter or other logic gate, one must also consider the possibility of avalanche breakdown of either an $N^+$ or $P^+$ source/drain diffusion or the punchthrough of one $N^+$ (or $P^+$) diffusion to another. This is because the output node, consisting of at least one $N^+$ diffusion wired to at least one $P^+$ diffusion, is either at the most positive voltage (in which case the full voltage is dropped across an $N^+$ diffusion) or at ground (in which case the full voltage is dropped across a $P^+$ diffusion within the N-well). Generally, logic gates are not designed to withstand a large amount of supply overvoltage. VLSI technologies in particular tend to have gated diffusion breakdowns less than a factor of two above the rated power supply voltage, especially for the $N^+$ diffusion. To avoid this complication during characterization, the 4 terminal PNPN structure, which has no FET gates and whose only junction required to withstand a large reverse bias is the N-well to substrate junction, is commonly employed for overvoltage stressing. This facilitates identification of the physical triggering mechanism, but is not necessarily representative of behavior on actual logic gates operating much higher than rated voltages.

A common display of the resulting characteristic was shown in Figure 5-2 on page 121. A less common, but more revealing, display results from a logarithmic plot of current vs. voltage. Often the detailed behavior of the blocking state characteristics

# Latchup Characterization

can be used to determine the triggering mechanism (e.g., punchthrough, avalanche, or field FET).

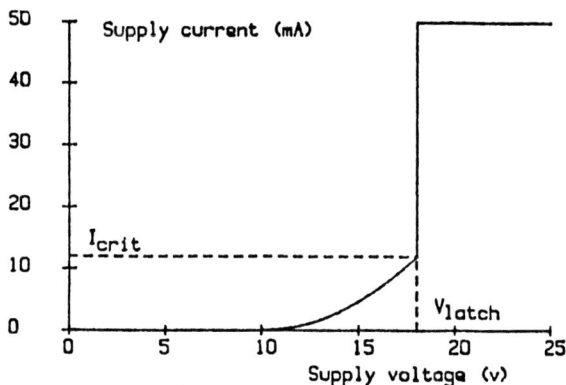

Figure 5-6. Supply overvoltage latchup characteristics. P-well depth is 0.8 μm, N-epitaxial layer is 5 μm, and $R_{sx}$ = 30 Ω. ($R_x$=0) From [Lewis-84]. © 1984 IEEE. Reprinted with permission.

In the absence of an equivalent source resistor ($R_x = 0$), one does not observe a negative differential resistance. Figure 5-6 shows supply current vs. increasing supply voltage for a PNPN structure [Lewis-84]. When the supply current reaches the switching current $I_s$, positive feedback reaches unity, and supply current rises rapidly to the compliance limit set for the power supply. The supply voltage at which the rapid rise occurs is $V_{latch}$, also referred to as the switching voltage or trigger voltage, and this voltage does depend on the triggering mechanism.

## 5.2.2 Supply Overcurrent Stress

One slight problem with overvoltage stress when using the parameter analyzer is that part of the I/V curve is obscured by the load line when switching from the blocking to the latched state. Since the characteristic is single valued in current, a possible alternative is to use a current source for the supply.

Although the measurement using overcurrent stress seems relatively simple, a potential problem exists which can have destructive effects on the CMOS structure being measured [Leventhal-84]. To turn on one of the parasitic transistors, a voltage drop across either the well or substrate bypass resistance is required to forward bias the corresponding base/emitter junction. If this lumped resistance is equal to 1 k$\Omega$, current approximately equal to 500 $\mu$A would be necessary to turn on the transistor, and this current would have to flow across the well/substrate junction. The reverse leakage current of this junction is typically under 100 pA, so some other phenomenon, such as avalanche breakdown, internal or external punchthrough, or parasitic FET action under the recessed oxide, would have to occur before the necessary current level is reached. If the voltage at which one of these effects takes place is high, device burnout may occur due to high power levels.

To facilitate latchup, Leventhal investigated the addition of external resistance to the well or substrate contact and the addition of an offset voltage between the base/emitter terminals of one transistor. However, unrealistically large resistances (in the G$\Omega$ range or higher) are required in the first case. In the second he successfully exhibited the complete I/V characteristic, but the trigger voltage is sensitively dependent on the offset voltage.

## 5.3 Three and Four Terminal Characterization

In supply overvoltage and overcurrent characterization, the PNPN structure is treated as a two terminal device, and latchup is said to be internally triggered. When avalanche, punchthrough, and field threshold voltages are all large, two terminal characterization of the blocking state yields little useful information, necessitating more extensive characterization. Just because a reasonable range of supply overvoltage fails to trigger latchup does not preclude other triggering modes, such as I/O over- or undershoot.

If external gating is used to trigger latchup, at least three terminals must be presented to the measuring apparatus to study the blocking state. One such example was shown in Figure 5-1(c) on page 119. Four terminal characterization, made possible by the parameter analyzer, simultaneously displays the bypass and emitter currents of both transistors, which is especially helpful in elucidating the sequence of events leading to latchup.

# Latchup Characterization

Consequently, this section addresses four terminal characterization of a PNPN structure using the parameter analyzer.

When the four terminals of the PNPN device are left individually accessible, latchup can be initiated by connecting the NPN emitter and its bypass resistor to ground, connecting the PNP emitter and its bypass resistor to a fixed $V_{dd}$, and then properly changing the voltage or current at one of these terminals. Since a voltage or current source can be applied to each of four terminals, there are a total of eight basic ways of externally gating latchup using a single supply. While some of these measurement techniques have limited applications, others can be used for detailed characterization of both the blocking and latched states[1].

### 5.3.1 Voltage Excitation at External Emitter or Base Terminal

One possible measurement technique applies an external base-emitter voltage to turn on one of the transistors. For example, the VPNP could be excited by connecting the external $N^+$ and substrate terminals to -5 volts, grounding the well terminal, and raising the internal $P^+$ voltage above zero. However, there is a problem with this technique. When the $P^+$ voltage is zero, the only current in the structure is the reverse leakage current of the well/substrate junction, so the well current starts positive. As the $P^+$ voltage increases, the emitter/base junction of the PNP becomes forward biased, initiating transistor action. For low values of $P^+$ voltage the NPN remains essentially off because of its miniscule base/emitter voltage. Any current flowing out of the PNP base must travel through the well contact, since the well junction passes only leakage current. Once the base current exceeds this leakage current, the well current becomes negative. For larger $P^+$ voltages, sufficient PNP collector current flows through the substrate resistance to forward bias the NPN base/emitter junction, in turn causing the NPN collector to begin draining some of the PNP base current. Eventually the NPN collector will demand more current than the PNP base can supply,

---

[1] See [Leventhal-84] for additional material on these measurement techniques.

so the remaining current will have to be supplied by the well terminal. At this point the well current will switch direction a second time and begin to flow into the structure. Since the well is grounded, the internal base voltage of the PNP will drop below ground (assuming $R_w \neq 0$). In order to maintain a smoothly changing forward bias voltage, the emitter voltage should drop with the base voltage. However, if it is being ramped constantly upward, the emitter voltage cannot respond, and an abrupt change in base/emitter bias occurs.

There are now two separate excitations driving the PNPN toward latchup. Since the voltage source on the emitter cannot adjust to changing base voltage, it no longer controls the excitation. Even if the emitter voltage were held constant, base drive would cause PNPN current to increase to the switching current, resulting in latchup. Thus, latchup appears to occur once the bypass current has reversed. In fact, the fully bypassed PNPN would not latch until the reversed current had increased enough to provide the full emitter/base forward bias.

One of the two excitations is absent when $R_w = 0$, which provides an interesting case for discussion. In this case the relevant form of the differential latchup criterion is identical to that for a floating N-well, since the PNP is not operating in a bypassed mode. Latchup occurs when the VPNP collector has turned on the LNPN so that its small-signal emitter resistance has been reduced to

$$r_{en} = \frac{R_s(\alpha_{fns} + \alpha_{fps} - 1)}{(1 - \alpha_{fps})}. \qquad (5-1)$$

In all likelihood the well current has already reversed direction for the second time when the above condition obtains. When the voltage source is returned to zero, the PNPN structure returns to the blocking state. It should be obvious that zero bypass resistance guarantees the PNPN tetrode configuration cannot latch, and latchup occurs only so long as the VPNP emitter voltage is applied. A voltage source on the emitter pre-disposes the PNPN to latchup, an effect exactly opposite to using a substrate or well reverse bias to harden the PNPN against latchup. Latchup is observed for this particular condition, not for the unbiased tetrode.

If the behavior of the circuit is analyzed when a voltage source is connected to any of the other three terminals, a similar problem is discovered. This measurement is useful only for studying behavior in a modified PNPN configuration.

An additional difficulty with voltage source triggering using the parameter analyzer is that incrementing voltage by a fixed amount causes exponential current changes. This reduces the measurement accuracy of the switching point for even the PNPN configuration.

### 5.3.2 Current Source Excitation on Bypass Resistor Terminal

Another possible measurement technique is to connect a current source to either the well or substrate terminal, while maintaining a voltage $V_{dd}$ across the structure. The direction of the current source is chosen to turn on the corresponding transistor by base current excitation. This measurement has been frequently used for SCR characterization and was discussed in section 5.1.1.

Such a connection alters the tetrode configuration to a triode, since the first transistor's base must source (or sink) the second transistor's collector current as well as current to the current source. Switching currents measured in this way are not generally representative of the tetrode configuration. It is a reasonable estimate only if the first transistor's bypass resistance greatly exceeds the second, i.e., it is the second transistor and its bypass resistor that chiefly determines the differential latchup criterion. Then, the transistor most affecting the latchup criterion is operating in the bypassed mode at the switching point.

Such an example is shown in Figure 5-7. This data is from the same sample as that used for supply overvoltage measurements illustrated in Figure 5-6 and was measured using a current source on the P-well [Lewis-84]. Total bypass resistance for the VNPN exceeded that for the LPNP so the triode configuration is a good approximation to the tetrode in this instance. In fact, the inverse switching current varied linearly for $R_{sx}$ from 0 to 100 Ω (in series with $R_{si} = 20$ Ω), from which it can be concluded that $R_w > 500$ Ω. As long as the PNPN configuration is not altered by test conditions, switching current is independent of the triggering mechanism. Note the switching current (designated $I_{crit}$ in this reference) is 12 mA in both these figures.

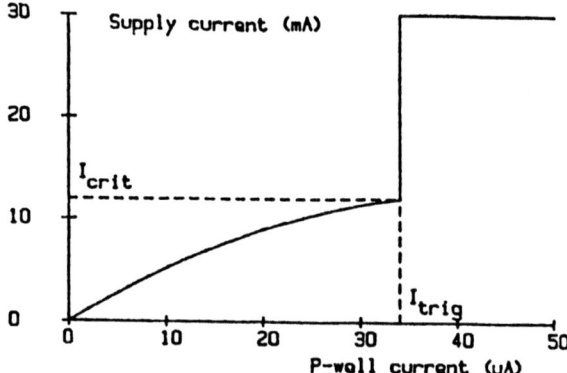

Figure 5-7. P-well current triggering. Supply overvoltage triggering. $V_{dd} = 5$ V. P-well depth is 0.8 μm, N-epitaxial layer is 5 μm, $R_{sx} = 30$ Ω and $R_x = 0$. From [Lewis-84]. © 1984 IEEE. Reprinted with permission.

Since the sign of the current cannot be changed on the parameter analyzer while performing a logarithmic sweep, a negative current sweep done to excite the PNP base current will display neither the region dominated by leakage current nor the portion of the blocking state for which the bypass current has reversed. Due to the bypass current's double-valued behavior, a monotonically changing current source will only be useful for measuring points up to the peak negative value. Since this peak can occur before the negative differential resistance region is entered, the tetrode's switching point cannot be accurately measured in all cases. Care must be taken not to mistake the last measurable value for the switching point, since the anode voltage has not yet reached its maximum. Similar double-valued behavior will be observed for the substrate current when trying to excite the lateral NPN, so the same limitations will be present if the current source is connected to the substrate terminal.

### 5.3.3 Current Source Excitation on External Emitter Terminal

In order to trace out the complete behavior of the PNPN structure as it switches from the blocking state to the latched state, the variable that is sourced should maintain control of the switching process and should exhibit monotonic dependence on total current. The voltages on any terminal fail the first stipulation, and the currents into the well or substrate terminal fail the second. Using a current source in the emitter maintains control of this transistor's turn-on since the emitter voltage can adjust to changing bypass current. The emitter/base forward bias is determined strictly by the value of emitter current. The current into either emitter terminal is monotonic over the entire curve from the blocking state through the negative differential resistance region and into the low impedance region. Since the transitors are off throughout most of the blocking state, the emitter current starts at zero, increases monotonically as the transistors are turned on, and continues to increase as the transistors are driven into saturation.

An example for a parasitic N-well PNPN structure is shown in Figure 5-8, in which the emitter current of a VPNP is sourced with the substrate and external $N^+$ diffusion held to -1 volts. The emitter voltage $V_{ep}$ and terminal currents $I_{rw}$ and $I_{en}$ are measured. The display in Figure 5-8 was chosen to exhibit the bipolar turn-on characteristics while providing a rough measurement of the critical currents and voltages. Because the only sink for base current is through $R_w$ when the VPNP first turns on, the VPNP base voltage is slightly positive, so that $V_{ep}$ too must be positive. (Current is defined to be positive when flowing into a sample's terminal.) The LNPN is out of the picture until $V_{ep} \simeq 0.7$ V, then its emitter current $I_{en}$ rises exponentially as current through the bypass resistor $R_s$ continues to force it on. At $V_{ep} \simeq 0.85$ V the substrate current $I_{rs} \simeq 0.5$ mA across $R_s \simeq 1$ kΩ produces a forward emitter/base bias $V_{en} \simeq 0.5$ V and an emitter current $I_{en} \simeq 0.6$ μA. Subsequent increases in substrate current produce a very sharp rise in LNPN emitter current $I_{en}$ because of the rapidly increasing emitter/base drive. Once the LNPN collector current can sink the VPNP base current, $I_{rw}$ reverses direction. The LNPN emitter current $I_{en}$ then rises sharply to the pre-set compliance limit, as does $I_{rw}$.

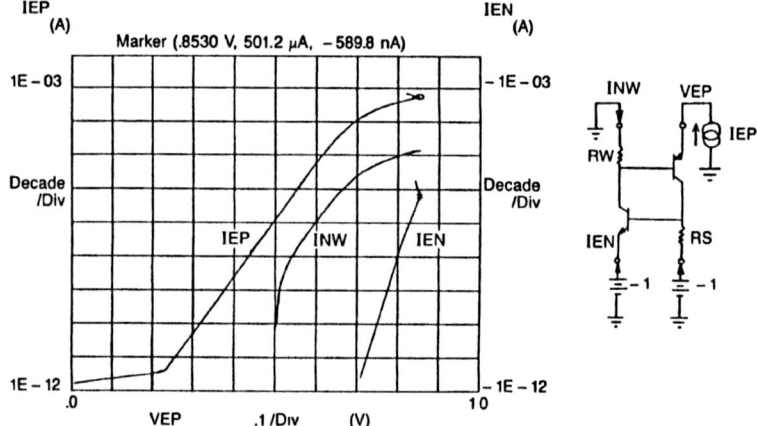

Figure 5-8. Switching from the blocking state. The VPNP is forced on using a current source on its emitter. The LNPN is eventually turned on by current through the bypass resistor $R_s$.

The critical point at which switching from the blocking state occurs is known as the switching point and is shown in more detail on the linear scale of Figure 5-9(a). This is a more accurate measurement than that shown for the previous logarithmic scale because the data points can be placed closer together in the vicinity of the switching point. It is also a more accurate measurement of both zero differential resistance points than is possible with the curve tracer. I is the total current into the parasitic PNPN device, and the open circle marks the switching current, here seen to be 0.55 mA. Also shown is the emitter current for the NPN which is just beginning to turn on. From these curves it is clear that NPN bypass current greatly exceeds emitter current at the switching point. Note that a portion of the negative differential resistance region is measurable before the system becomes unstable.

Analagous characterization is possible for the holding point, i.e., the point in the latched state at which the PNPN's differential resistance is zero. Figure 5-9(b) illustrates this region using an expanded scale for the terminal voltage V. These curves were made by first switching the device into the low impedance state and then using a current source on the PNPN tetrode configuration to reduce total current through the device. Note the holding current of 1.5 mA is nearly three times the switching current.

(a)

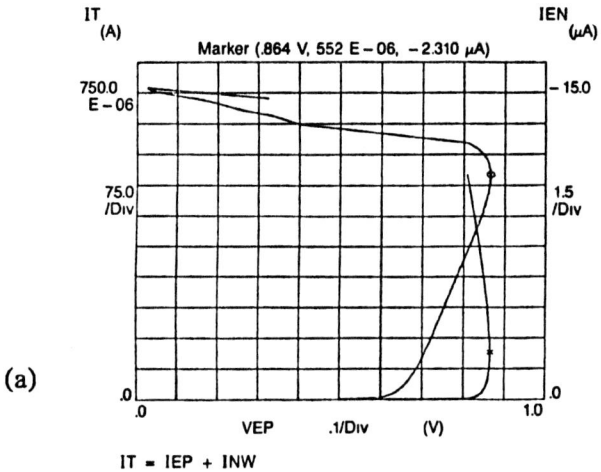

(b)

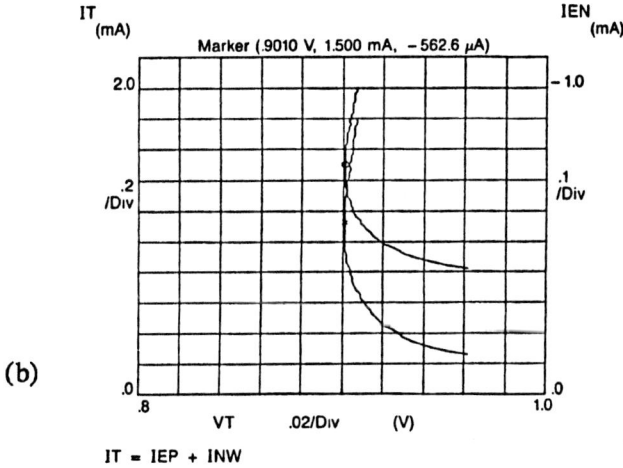

Figure 5-9. Points of zero differential resistance. (a) Switching point and (b) holding point. I is the total current into the device, and $I_{en}$ is the second transistor emitter current.

In the next section, current source excitation on the emitter terminal is used to study the dependence of switching current on bypass resistance. The data presented is in excellent agreement with the differential latchup criterion derived in Chapter 4.

## 5.4 Switching Point Characterization

Emitter current excitation methods are very useful for blocking state characterization because switching conditions are experimentally well controlled and the switching point is well defined. In most cases part of the negative differential resistance region also will be measurable before the structure becomes unstable.

Sample curves are shown in Figure 5-9(a) for the case when the current source drives the PNP emitter. The measuring circuit is shown in Figure 5-8. As emitter current $I_{ep}$ forces on the PNP, base current flowing through $R_w$ produces a positive base voltage and a base/emitter forward bias commensurate with $I_{ep}$. Consequently, the PNP emitter voltage $V_{ep}$ changes with changing emitter current. It is relatively easy to derive the condition

$$\frac{dV_{ep}}{dI} = \frac{[1 - (\alpha^*_{fns} + \alpha^*_{fps})]R_{w1}}{\alpha^*_{fps}}, \qquad (5-2)$$

so that the differential latchup criterion for the PNPN tetrode configuration holds when this derivative is zero. The digital readout of the parameter analyzer's marker can be used to easily determine the currents at which $dV_{ep}/dI = 0$.

Both total current $I = I_{ep} + I_{rw}$ and NPN emitter current $I_{en}$ are included in this plot. The larger of these two critical currents is the switching current, and it depends in a known and predictable manner on the type of triggering used. The smaller current is the second transistor's (the NPN in this case) emitter current at the switching point. (The first transistor has already been turned on by the current source at its emitter.) Both critical currents have been measured over a wide range of external bypass resistances to test the behavior predicted by the differential latchup criterion. To facilitate this comparison, the external resistors can be chosen

# Latchup Characterization

so that triggering is well described by the triode approximation to the general criterion.

When the current source drives the PNP emitter, for example, the external resistor values are chosen such that $R_w >> R_s$. This insures that NPN turn-on controls the latchup sequence, as it does for the floating N-well case. Once NPN base/emitter voltage reaches the value predicted by equation (4-57), drive on the PNP resulting from NPN collector current through the bypass resistor dominates the current source drive, the PNPN device enters the negative differential resistance region, and PNP emitter voltage falls to reflect the polarity reversal of the N-well current.[2]

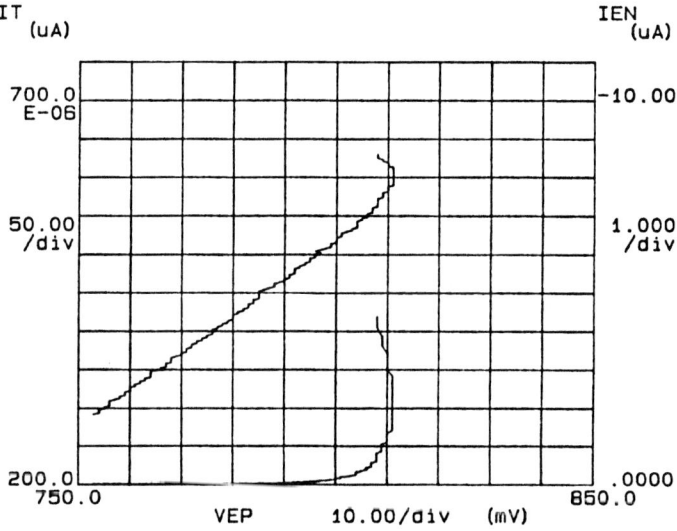

Figure 5-10. Illustration of 1 mV measurement resolution. Upper curve is total PNPN device current; lower curve is second transistor emitter current.

---

[2] In [Takacs-83] supply overvoltage is used to measure switching current (called critical current in this reference) as a function of voltage on a gate electrode straddling the N-well junction from the internal $P^+$ to external $N^+$ diffusion. Switching current decreases with increasing $R_w$ for the non-epi case but not for the epi case. In the epi case $R_s << R_w$ so that $R_s$ chiefly determines switching current, while for the non-epi case $R_s > R_w$, and $R_w$ chiefly determines switching current.

Although a current source on the emitter is well suited to measurements in the vicinity of an inherently unstable operating point, the 1 mV measurement resolution of the parameter analyzer does limit the accuracy of determining the precise current at which $dV_{ep}/dI = 0$. This is illustrated on the expanded voltage scale of Figure 5-10 for both total current $I_{ep} + I_{rw}$ and NPN emitter current. Also, repeated measurements do not always yield exactly the same result. In subsequent figures showing this data, a vertical bar represents the full range of three consecutive measurements.

Knowledge of several transistor parameters is required for a detailed comparison with the model in Chapter 4. Figure 5-11 shows small-signal alphas and betas as measured for the devices in the present comparison using the appropriate three PNPN terminals. Emitter current is sourced, and the collector/base reverse bias is 5 V. Subsequent measurements of switching current are made for a range of bypass resistance values, and the range of emitter currents for these measurements are marked on the current gain curves both when the transistor is the first transistor (being excited) and when it is the second transistor to turn on. Clearly, high level injection effects can be neglected for the second transistor. In addition, from the logarithmic slope of emitter current vs. emitter voltage, the equivalent thermal voltages measured 26.4 and 26.1 mV for the NPN and PNP, respectively, and saturation currents for the base/emitter junctions measured $I_{sn} = 7.9$ fA and $I_{sp} = 0.33$ fA.

Internal substrate resistance is most easily deduced by plotting inverse switching current vs. external substrate resistance. An example is shown in Figure 5-12 for external substrate resistors from zero to 1 kΩ. The solid curve has been calculated using equations (4-56) and (4-58), the previously mentioned parameter values, and an internal substrate resistance of $R_{si} = 850$ Ω. This is essentially the value calculated for the resistance from the small area topside substrate contact to the highly doped substrate below the epitaxial layer. Series base and emitter resistance are negligibly small and have been omitted from the calculations. Note the calculated curve is slightly super linear, and linearly extrapolating the data would not yield quite the same value.

# Latchup Characterization

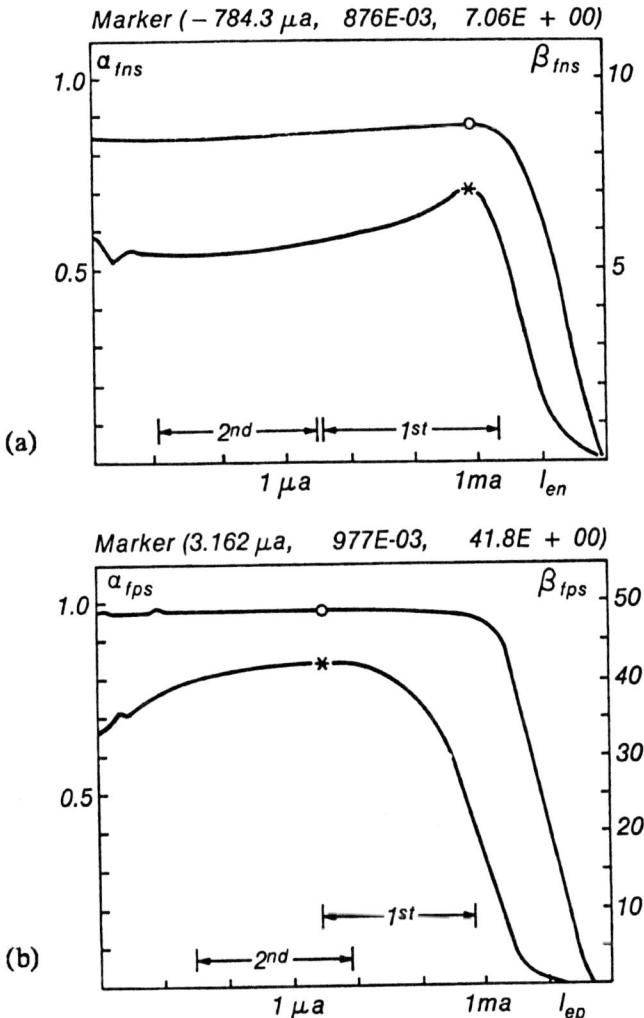

**Figure 5-11.** Small-signal alpha and beta. (a) LNPN and (b) VPNP. Emitter current is sourced and collector voltage is 5 volts in both cases.

Similar behavior has been reported for a PNPN structure fabricated with a shallow (0.8 μm) P-well in a 5 μm epitaxial layer over an $N^+$ substrate [Lewis-84]. A backside contact was used, and switching current (called critical current in this reference) was measured by raising the power supply until latchup occurred, in this instance triggered by punchthrough from the substrate to the internal $N^+$ diffusion. Inverse switching current increased linearly as external substrate resistance was varied from 0 to 100 Ω, and the extrapolated internal substrate resistance was 20 Ω. The forward bias on the lateral emitter/base junction at the switching point was deduced from the same curve to be 0.58 V.

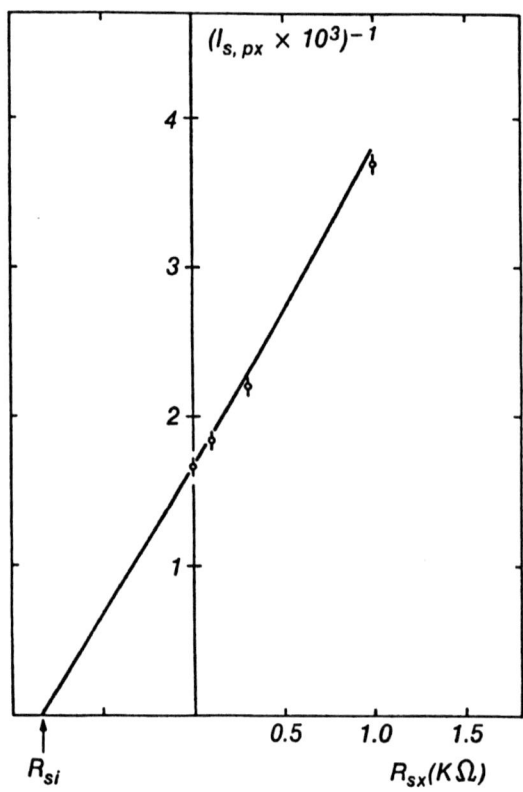

Figure 5-12. Determination of internal substrate resistance. $I_{sp,x}$ is the switching current measured when the PNP emitter current is sourced. $R_{sx}$ is external NPN bypass resistance; $R_{st}$ = 850 Ω.

Figure 5-13 compares predicted and measured switching currents over a wider range of substrate resistance. Switching current is measured with the current source on the PNP emitter, and it is plotted vs. the inverse substrate resistance in milli-Siemens. A value of 850 Ω has been added to the external substrate resistance to obtain total substrate resistance $R_s$. Bypass

# Latchup Characterization

resistors have been chosen so that $R_w >> R_s$ for all data. This choice guarantees the triode configuration with $R_w$ removed is an excellent approximation to the tetrode, and it facilitates data comparison to the model derived in Chapter 4. The top curve has been calculated from (4-56) and (4-58) using $\beta_{fps} = 20$, corresponding to the emitter current for $R_{sx} = 0$, and the bottom curve using $\beta_{fps} = 42$, which is the appropriate value for $R_{sx} = 100$ k$\Omega$. Because the curves are so close together, no attempt has been made to calculate the full curve using intermediate $\beta_{fps}$ values at each emitter current. One can see, however, the trend in the data from the lower to the upper curve as $R_{sx}$ is decreased. Also, note the predicted curves are well approximated by a straight line on this log/log scale but that the slope is 1.16 rather than unity because the critical base/emitter voltage is a function of bypass resistance.

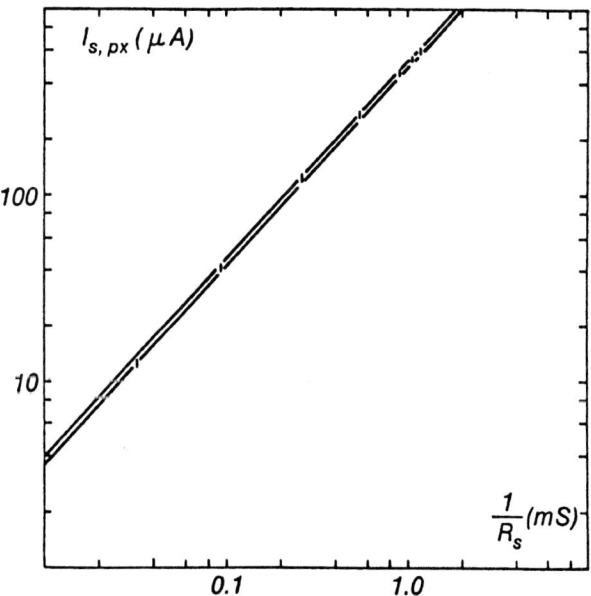

Figure 5-13. Switching current for PNP-driven triode configuration. The vertical bars represent the experimental range of 3 measurements. The upper curve is calculated using $\beta_{fps} = 20$, the lower, $\beta_{fps} = 42$.

The driving transistor's beta dependence on emitter current has a much more pronounced effect on the second transistor's emitter current at the switching point than on the switching current. Figure 5-14 clearly shows this behavior over the same range of external bypass resistance discussed above. Now the curve has been calculated from (4-56) using the appropriate values of gain from Figure 5-11 on page 141. The beta falloff raises the second transistor's emitter current at the switching point for the smaller $R_s$ values. In spite of the greater data spread, the observed current is seen to follow the predicted behavior reasonably well. Theoretically, if the internal substrate resistance were small enough, the curve would asymptotically approach infinity at that value of substrate resistance for which the beta product is unity.

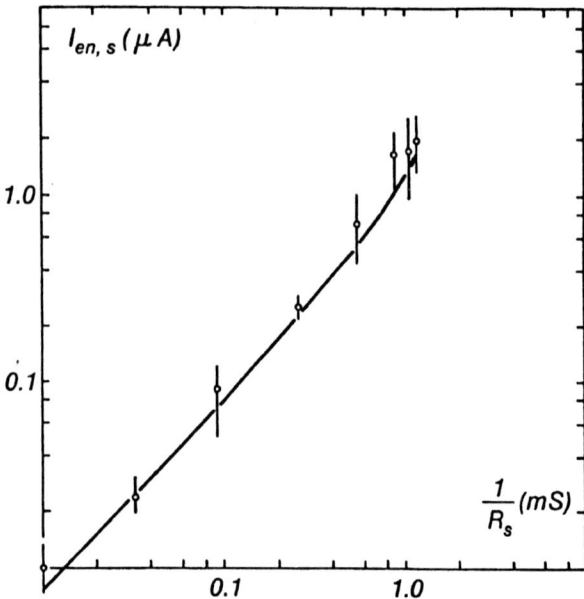

Figure 5-14. Second transistor emitter current at switching point. PNP-driven triode configuration. $R_s$ is the total NPN bypass resistance.

Measurement of latchup controlled by the PNP can be made by placing the current source on the NPN emitter and choosing external bypass resistors such that $R_s >> R_w$. Figure 5-15 compares measured switching current with the floating substrate

## Latchup Characterization

predictions of equations (4-59) and (4-61). This curve yields an internal well resistance of 285 Ω and clearly shows the slightly super-linear dependence of inverse switching current on total well resistance resulting from the dependence of critical forward bias on bypass resistance. The excellent agreement between measured switching current and the predictions of the differential latchup criterion is also evident in Figure 5-16, which compares the two for well resistances spanning 2.5 orders of magnitude.

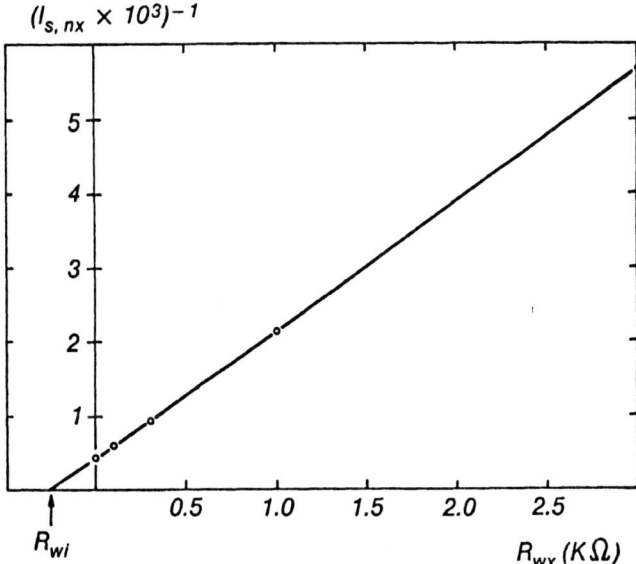

Figure 5-15. Determination of internal N-well resistance. $I_{sn,x}$ is the switching current measured when NPN emitter current is sourced. $R_{wx}$ is external PNP bypass resistance; $R_{wi} = 285$ Ω.

Comparison of measured and modeled emitter current when the PNP is the second transistor reveals an important additional feature. As noted above, the second transistor's emitter current is a much stronger function of the driving transistor's current gain than is the switching current. When the modeled results use the standard 3-terminal NPN current gain curve (measured with the PNP emitter floating), predicted emitter current at the switching point is higher than that actually measured for the smaller external well resistances, as can be seen in Figure 5-17. In the standard measurement, base current is supplied only by the substrate contact, but when the VPNP is involved in cooperative action with the LNPN, holes are supplied to the lateral base from the VPNP as well. Holes reaching the substrate from the VPNP and exiting from a topside substrate contact establish an electric field in the substrate, which enhances the N-well's collection of injected electrons and raises current gain. Estreich has studied field-aiding effects in a lateral transistor and has shown that even for a relatively small electric field (10 to 100 V/cm), the lateral beta is substantially enhanced when base width exceeds the minority carrier diffusion length [Estreich-80].

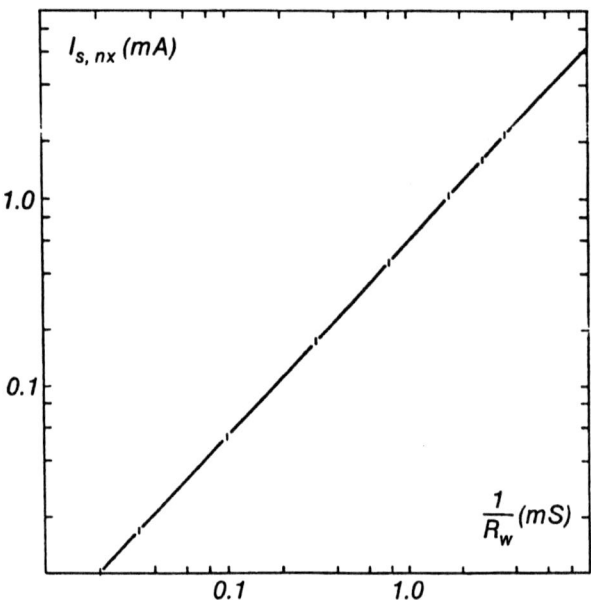

Figure 5-16. Switching current for NPN-driven triode configuration. The vertical bars represent the range of 3 measurements. There is no perceptible difference in the calculated curve using $\beta_{fns}$ values in the range of 5.8 to 7.

Several authors have noted higher lateral gains when holes are supplied only by the VPNP. (See [Ochoa-81] and [Leventhal-84] for examples.) Such a measurement entails floating the substrate and sourcing VPNP emitter current to supply NPN base current. Since both transistors are active, latchup occurs for emitter currents exceeding some critical level, and this technique does not

# Latchup Characterization

provide the entire beta curve. The dashed curve in Figure 5-17 is calculated using $\beta_{fps}$ measured by this technique. Note that it tends to the low side of the data.

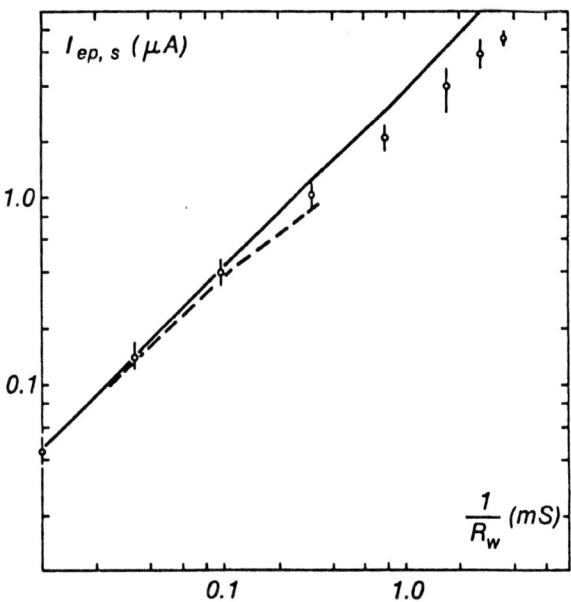

Figure 5-17. Second transistor emitter current at the switching point. NPN-driven triode configuration. $R_w$ is the total PNP bypass resistance. Solid curve - holes supplied only by topside substrate contact for current gain neasurement; dashed curve - holes supplied only by VPNP.

Alternatively, if the emitter current at the switching point is used to deduce lateral small-signal beta, the solid curve of Figure 5-18 results. For comparison, the dashed curves illustrate small-signal behavior when holes are supplied by substrate contact only (standard measurement) and by VPNP only. The curve deduced from latchup measurements falls between the other two, suggesting that both groups of holes play an important role in determining the switching point. Because VPNP bypass current greatly exceeds its emitter current at the switching point, this sensitivity has negligible effect on the switching current plotted in Figure 5-16.

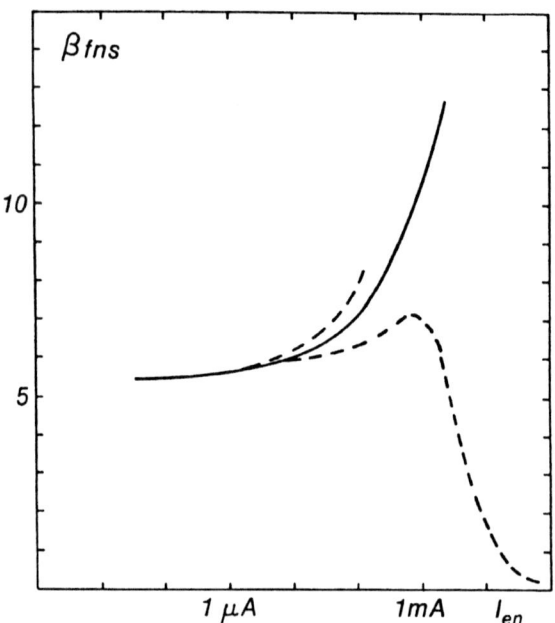

**Figure 5-18.** Comparison of lateral, small-signal betas. In the lower dashed curve NPN base current is applied by substrate contact only (standard 3-terminal measurement); in the upper, base current is supplied by the VPNP only. The solid curve is deduced from the $I_{ep,s}$ measurement shown in previous figure.

In summary, for the same value of bypass resistance, the second transistor's current at the switching point is larger when the VPNP controls the latchup sequence because the LNPN, as the driving transistor, has lower small-signal beta. The ratio $I_{ep,s}/I_{en,s}$ is approximately 5x for the larger external bypass resistances but diminishes for smaller bypass resistance because $\beta_{fps}$ falls off at higher emitter current more than does $\beta_{fns}$. Everything else being equal, one would then expect switching current to be larger when the VPNP controls latchup just due to the small-signal beta difference. However, PNP emitter/base saturation current is 24 times less than that for the NPN, which causes the difference to be even greater. In fact, the difference for equal bypass resistors $R_w = R_s = R$ is easily shown to be approximated by

$$I_{sn,x} - I_{sp,x} = \frac{V_t}{R}[\ln(\frac{\beta_{fps}I_{sn}}{\beta_{fns}I_{sp}})] \ . \tag{5-3}$$

Clearly, any process or layout design undertaken to avoid latchup must consider the lower of the two.

Finally, we note the differential latchup criterion derived in Chapter 4 is in excellent agreement with measured switching point data. This agreement extends over many orders of magnitude for bypass resistance, making the differential latchup criterion an accurate predictor of switching current for any layout. There are no fitting parameters since the critical value for emitter/base forward bias is precisely determined by the differential latchup criterion.

## 5.5 Holding Point Characterization

The low impedance state is studied by first triggering the PNPN structure into latchup, by applying a base/emitter voltage for example, and then removing the triggering source while maintaining device current above the holding current, preferably with a current source. The measurement set-up is then the intrinsic two terminal configuration, and measurements of the low impedance region can be made by decreasing the current.[3] Care must be taken to limit the total voltage across the device, as lowering the current will eventually result in a traversal through the negative resistance region back to the high impedance state. Since device burnout can result in the high impedance state when attempting an overcurrent stress, similar burnout can occur when switching back to this state if the voltage is not limited.

A current source is preferable when tracing down the latched state characteristic because the initial anode voltage could cause excessive anode current to flow, but as long as this current is

---

[3] Simply raising the power supply voltage until latchup occurs and then measuring the lowest current for the latched state does not usually yield the holding voltage. Such a point is determined by the dynamic load line.

somehow limited, a voltage source can be used without destroying the sample. If a voltage source is used, series resistance can be added to measure a portion of the negative differential resistance characteristic and clearly display the "knee" at which $dV/dI = 0$.

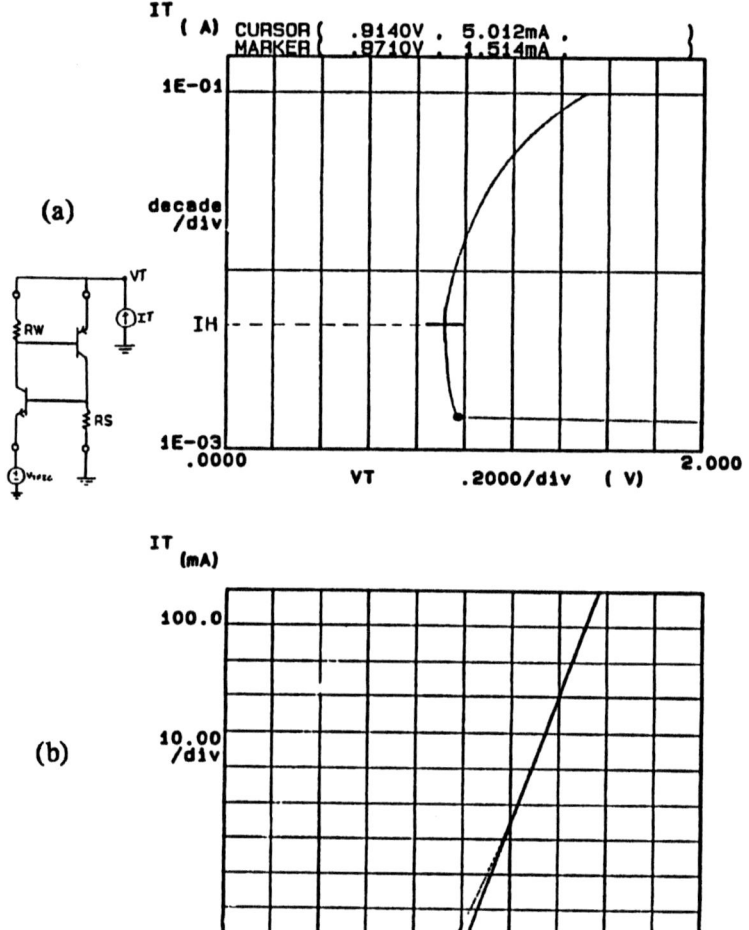

Figure 5-19. Latched state I/V characteristic. (a) Logarithmic scale. (b) Linear scale. From [Leventhal-84]. Reprinted with permission.

Plots of the low impedance state characteristic are shown in Figure 5-19. Part of the negative differential resistance region below the holding current is measurable, but eventually the instability becomes too large, and the voltage across the current source jumps out to the compliance limit. Since part of this region is measurable, the holding point is well defined,[4] and the values of holding voltage and current are easily identified and accurately measured. In the linear plot of Figure 5-19(b), the incremental resistance in the low impedance state is 5.6 Ω. The contact resistance of the probes was measured separately to be several ohms per tip, so the actual incremental resistance of the parasitic PNPN structure is extremely small.

Some interesting properties of the holding point have been demonstrated using a pair of discrete, cross-coupled transistors with lumped resistors for the resistances shown in the lumped element model [Dressendorfer-81]. Adding resistance to the emitter leg raises holding voltage and the latched state's resistance, but the I/V characteristic for the latched state still extrapolates to the origin. Adding collector resistance ($R_{s2}$ or $R_{w2}$) not only raises the holding voltage but also shifts the extrapolated latched state characteristic to a positive intercept on the voltage axis.

In [Fang-84] "holding current" is measured by disconnecting one pair of base/emitter terminals and using base current to trigger the PNPN into the latched state, then reconnecting the base and lowering anode voltage to find the minimum measurable current on the low impedance I/V characteristic. Data shows the minimum measurable current in these non-epi samples, which is slightly lower than the real holding current, to increase with increasing LNPN basewidth for both the normal and reversed PNPN structures. Sustaining current is higher for the reversed

---

[4] The holding point is defined here by $dV/dI = 0$. Some authors have taken holding current to be the lowest measurable current in the latched state (see, for examples, [Rung-83] and [Fang-84]), but, as is discussed in Appendix A, the measurable portion of the negative differential resistance region is measurement-dependent, so the lowest measurable current is not fundamentally significant. In [Rung-83] the difference between the holding point and the lowest measurable point exceeds 1 volt.

structure and decreases with increasing VPNP emitter to base distance (increasing $R_w$).

In [Hu-84b] the holding point is characterized by triggering latchup with intense light or by forward biasing PN junctions with the backside contact to the substrate disconnected. With the PNPN structure in the latched state the backside was reconnected, and the power supply voltage was slowly ramped down until the structure switched out of the latched state. Both holding current and holding voltage increased as the $N^+/P^+$ spacing in their 12 $\mu$m thick epitaxial layer samples was extended from 7 to 20 $\mu$m. Both increased still more when butted contacts were used in the N-well and in the substrate, although this increase was less dramatic for the closer spacings. A much more dramatic increase occurred when majority carrier guards were used in both well and substrate. The structure utilizing butted contacts also showed a dramatic increase in holding voltage when the final epitaxial layer thickness was reduced from 12 to 3.5 $\mu$m. For the 7 $\mu$m $N^+/P^+$ spacing, holding voltage exceeded 4.5 V and rose steeply with increasing separation. No data was presented for guarded structures on the thinner epitaxial layer.

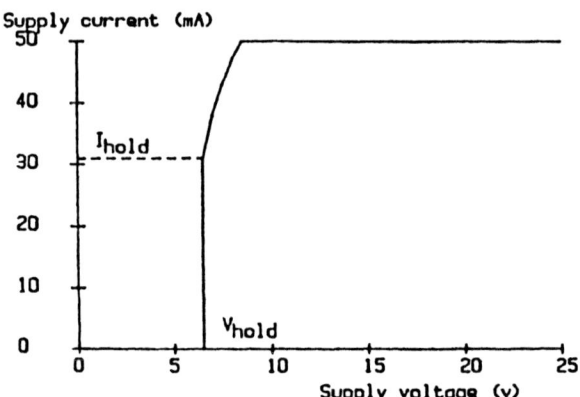

Figure 5-20. Holding point measurement using voltage source. The actual holding point is not clearly exhibited; $R_x = 0$. From [Lewis-84]. © 1984 IEEE. Reprinted with permission.

In [Lewis-84] latchup is triggered by supply overvoltage or by base current triggering. Then the supply voltage is lowered until

# Latchup Characterization

the PNPN device switches from the latched state, as shown in Figure 5-20. Note the zero differential resistance point is not displayed. Inverse holding current is shown to increase essentially linearly as external substrate resistance $R_{sx}$ is increased from 0 to 100 Ω. Extrapolating this curve to the intercept on the horizontal axis yields an internal substrate resistance of 10 Ω, which is half the value found by extrapolating inverse switching current, probably due to conductivity modulation. The corresponding holding voltage dropped from 22 to 2.3 V as $R_{sx}$ was increased.

In addition to studying the total current through the structure for the low impedance state, it is also interesting to look at the way total current splits between bypass and emitter currents. If the current source is first connected to the well and internal $P^+$ diffusion, the currents into the external $N^+$ and substrate terminals can be measured using the parameter analyzer. By switching the current source to the other two terminals, the corresponding currents into the well and internal $P^+$ diffusion can be measured. The results of these measurements can then be combined to illustrate the complete picture for the latched state. However, the characteristics from the two separate measurements should be compared at high current levels to insure they are identical. When the total current is equal to 100 mA, a 1 Ω difference in resistance, which is easily within the variability of probe contact resistance, will result in a 0.1 volt shift between the two curves.

Plots of the currents into the four terminals, graphed vs. the total current, are shown in Figure 5-21. For device current below the holding current, the NPN emitter current $I_{en}$ exceeds the corresponding bypass current $I_{rs}$, while the PNP emitter current $I_{ep}$ is less than its corresponding bypass current $I_{rw}$. As expected, the emitter currents dominate once the structure is well into the latched state. Deep into saturation the slopes of emitter current vs. total device current are as described by (4-79) and (4-80). Note the slopes would be unity if there were no reverse transistor action. Since these plots match up with the I/V characteristics shown in Figure 5-19, the holding current value is equal to 5.012 mA. Note the similarity to the behavior depicted in Figure 11 of Chapter 4, for which the parameter values are roughly equal to those of the measured device.

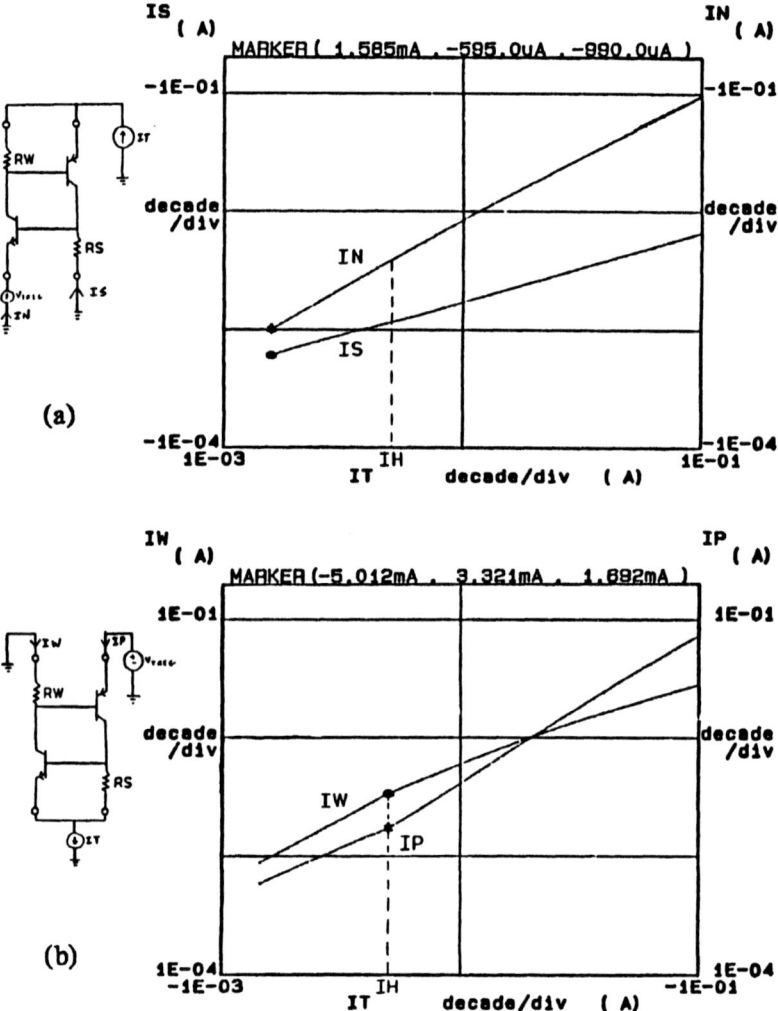

Figure 5-21. Low impedance state currents. Range shown includes the measurable negative differential resistance region. From [Leventhal-84]. Reprinted with permission.

## 5.6 Dynamic Triggering

The preceding sections of this chapter have all addressed latchup caused by various forms of static excitation. It is time to turn to transient excitation techniques, in which current flow through the latch structure is caused by some transient waveform. Here we seek the critical time rate of change or other property of the waveform that triggers latchup.

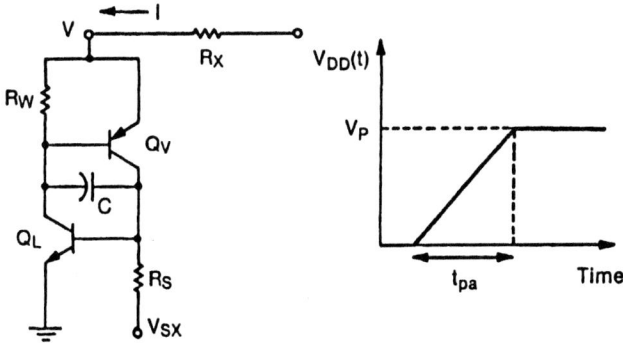

Figure 5-22. Transient excitation using power supply ramp. N-well junction capacitance causes displacement current to flow through bypass resistors. From [Troutman-83a]. © 1983 IEEE. Reprinted with permission.

One technique for providing the transient excitation is to ramp the power supply voltage from ground to its final value at a constant rate, as shown in Figure 5-22 [Troutman-83a]. The resulting displacement current flowing through the bypass resistors and the N-well junction capacitor can cause latchup if the ramp rate is above some critical level, even when the final voltage value is below that necessary for over voltage triggering. Fu has shown this critical ramp rate to be larger when the voltage dependence of this capacitance is included [Fu-85]. The factor can be as large as 2x and depends on the exact nature of the capacitance vs. voltage curve.

Figure 5-23 illustrates some of the data for latchup triggered by ramping power supply voltage [Troutman-84]. The vertical axis is the time required for the power supply to reach 5 V, and the height of each bar indicates the critical time before latchup occurs. Faster ramp rates can be tolerated by either applying a reverse bias to the substrate or reducing the series N-well resistance. Unguarded and three types of guarded structures were tested for both epi- and non-epi CMOS samples. The single most important feature providing protection against latchup was epi-CMOS with a backside substrate contact or, equivalently, with a large area topside contact. (See Figure 4 of [Troutman-86] for a comparison.) Even for the unguarded structure at zero substrate bias, latchup did not occur when ramp rates as fast as 1.7 V/nS were applied unless the series N-well resistance was on the order of 10 kΩ or larger. Non-epi samples using majority carrier guard structures in both N-well and substrate approach, but do not match, the latchup hardness of the epi-CMOS samples.

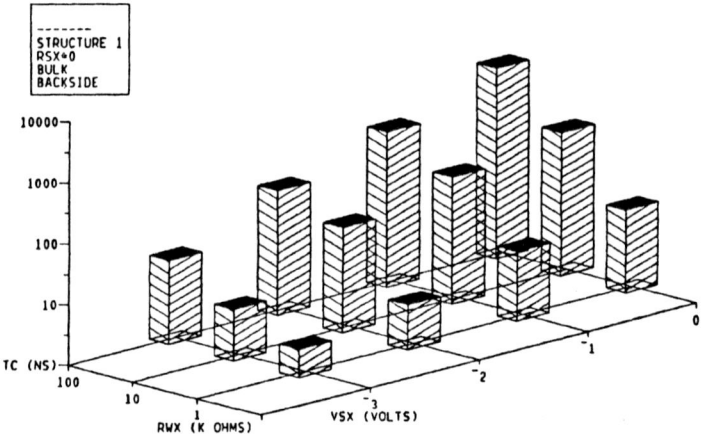

Figure 5-23.  Critical power supply rise time.  Unguarded PNPN structure on non-epi substrate with backside contact. The higher the bar, the greater the latchup sensitivity.  From [Troutman-84]. © 1984 IEEE. Reprinted with permission.

A second dynamic triggering technique applies a substrate current through two separate substrate contacts, which forces a current to flow through the LPNP bypass resistor [Rung-83]. Pulse amplitude and width are varied to determine the latchup conditions with the results shown in Figure 5-24. This curve has two asymptotic limits. A minimum pulse height is required no matter how wide the pulse. The switching current for this structure is 2.65 mA, and even DC current below this level does

## Latchup Characterization

not trigger latchup. This limit corresponds to the switching boundary of SAFE space discussed in Chapter 4. It is also clear that pulse widths below approximately 50 nS do not cause latchup, no matter how large the pulse height. Since the LPNP emitter/base voltage is applied directly using an external voltage supply, the response time in this case is principally due to the base transit times, which are longer than the RC times associated with the well/substrate junction. This technique of pulsing two separate substrate contacts is also reported in [Fang-84].

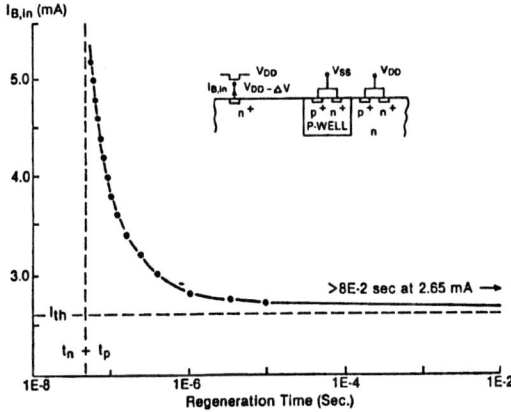

Figure 5-24. Transient excitation by base injection. Latchup is triggered by current pulse through LPNP bypass resistor. From [Rung-83]. © 1983 IEEE. Reprinted with permission.

In a variation of this technique, well current is applied using two separate contacts, which forces current through the bypass resistance of the VNPN [Goto-83]. Again pulse height and width were varied to find the pair of values causing latchup, and results are shown in Figure 5-25 for various spacings $W_1$ between internal and external $P^+$ diffusions. The asymptotic limit for wide pulses is approached for pulse widths greater than 0.4 $\mu$S, and this limit is reduced as the VNPN bypass resistance increases (either by longer distances to the well contact or reduced well conductance). Greater pulse height is required for shorter pulses, although this effect diminishes as $W_1$ is reduced, and latchup does not occur for pulse widths of 20 nS and below.

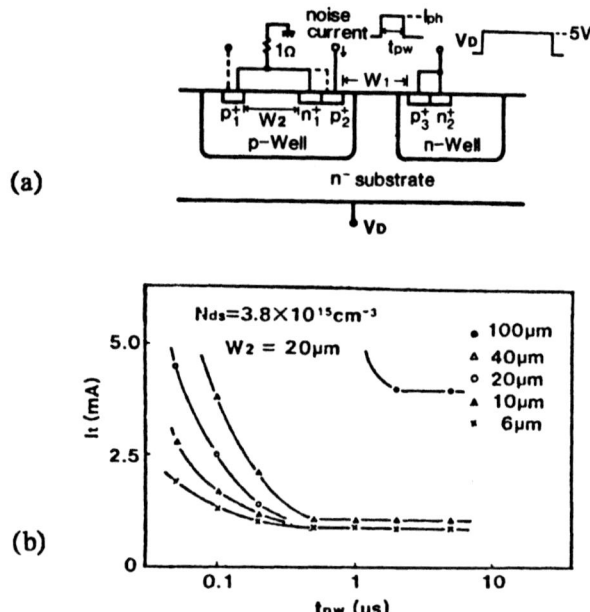

Figure 5-25. P-well Pulse Excitation. (a) Test Structure. (b) Pulse height vs. width required for latchup with $W_1$ as parameter. From [Goto-83]. © 1983 IEEE. Reprinted with permission.

In another pulse technique the pulse is applied directly to the available base and emitter contacts [Hamdy-83]. For LNPN excitation it is applied across the substrate and external $N^+$ diffusion while for VPNP excitation, across the N-well and internal $P^+$ diffusion. Because of the effective base resistance for this test configuration, the actual emitter forward bias is somewhat less than the applied voltage. As for representing tetrode latchup, this measurement configuration suffers the same difficulties discussed in section 5.3.1 for static excitation. However, it is intended to simulate I/O over- or undershoot conditions, for which an emitter bias actually does appear at these terminals.

The forward base/emitter bias approaches an asymptotic limit for wide pulses and rises sharply for narrow pulses, as shown in Figure 5-26. As pulse width is reduced, the voltage necessary for latchup increases until at 10 nS it is approximately 2X that when the LNPN is excited and 3X that when the VPNP is excited. N-well resistance for LNPN excitation makes it easier for the LNPN collector current to turn on the VPNP, so that less excitation is required. Raising temperature also reduces the required excitation due to an increase in N-well resistance and to an increase in the base/emitter reverse saturation current. The same general comments can be made about VPNP excitation, but now the dependence on temperature and the other transistor's bypass resistance is more pronounced because the VPNP has higher gain.

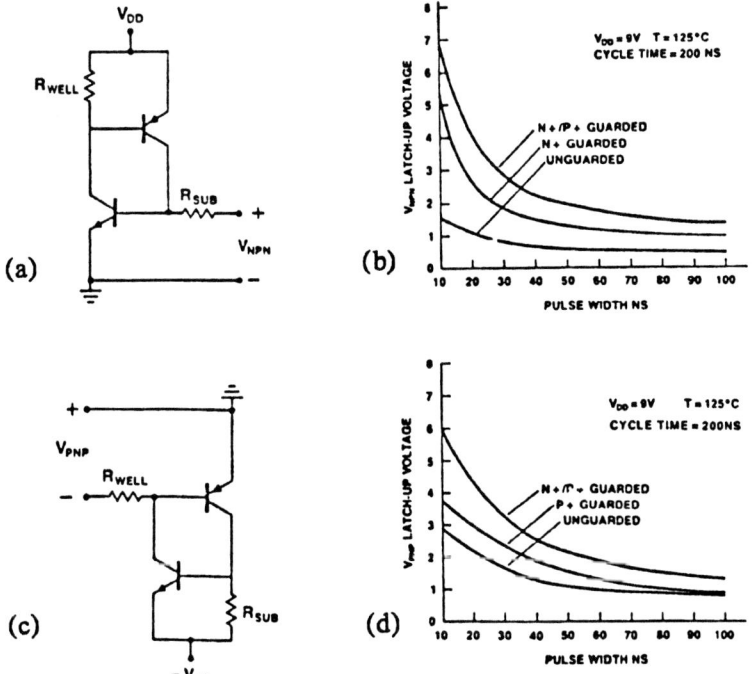

Figure 5-26. Emitter/base pulse excitation. (a) NPN initiated latchup. (b) NPN forward bias vs. pulse width with N-well resistance and temperature as parameters. (c) PNP initiated latchup. (d) PNP forward bias vs. pulse width with substrate resistance and temperature as parameters. From [Hamdy-83]. © 1983 IEEE. Reprinted with permission.

Cycle time was varied and no change was noted as long as the off time of the pulse was sufficiently long to allow injected base charge to recover. However, for cycle times less than 30 nS for NPN excitation and 50 nS for PNP excitation, the required base/emitter forward bias was observed to decrease. The amount of injected base charge then accumulated from one pulse to the next.

In addition, majority carrier guard rings raise the bias required for latchup by effectively reducing bypass resistance. Using both an $N^+$ guard ring in the N-well and a $P^+$ guard ring in the substrate was even more effective than using just one.

### 5.7 Temperature Dependence

Bypass resistance decreases as temperature is reduced, and the base/emitter forward bias necessary to support a given collector current increases. Both effects cause switching current and holding current to increase at lowered temperature. The current required to sustain latchup has been shown to increase by more than 2X when the temperature is reduced from 300 K to 77 K [Dooley-84]. When an inverter is triggered into latchup by sinking current at the output node (forcing on the parasitic VNPN), the trigger current increased by more than 4X for the same temperature reduction because of the additional temperature dependence associated with the $\beta$ of the output VNPN.

Injection-efficiency limited transistors typically exhibit an exponential falloff in common-emitter gain $\beta$ as temperature is reduced while transport limited transistors show a weaker dependence [Dooley-84]. Usually the vertical parasitic transistor is injection limited while the lateral parasitic is more likely to be transport limited. Consequently, $\beta$ for the vertical device is expected to exhibit greater temperature dependence.

Elevated temperature results show that both the VNPN common emitter current gain and the emitter/base bias on the LPNP at the holding point increase by approximately 2X between room temperature and 150 C [Lewis-84]. Holding current decreases by more than 2X. Trigger current (current is forced into the VNPN base) decreases by more than 5X because of the

# Latchup Characterization 161

additional increase with temperature of the VNPN beta. This data was taken using an external substrate resistance $R_{sx} >> R_{si}$ so that any temperature effects due to increased lattice scattering in the substrate are minimized. Heightened substrate resistance at elevated temperatures also reduces both holding current and trigger current.

## 5.8 Non-Electrical Probing

In addition to electrical probing via the pads of a clearly defined PNPN structure, there are non-electrical probing techniques that are especially helpful in determining latchup paths in complicated circuits or even entire chips.

One such technique uses a scanning electron microscope (SEM) in the electron beam induced current (EBIC) mode in which an image is formed from the current produced by collecting electron-beam generated electron-hole pairs in space charge regions [Dressendorfer-80]. Charge carrier collection depends on junction width and voltage and on minority carrier diffusion length. Because latched junctions are biased differently than unlatched, and because minority carrier density can be significantly altered by high level injection in the latched state, the EBIC signal from latched regions is measurably different from those not involved in latchup. In measurements made on a 1 Kb RAM chip, latchup was excited by overvoltage, and latched areas appear brightest in the picture. Superposing a regular SEM picture of the area around the latched portion with the EBIC image helps to identify junctions participating in latchup. Different paths occurred on different chips, and even on the same chip retesting sometimes revealed a different set. The authors attributed this inconsistency to variability of the internal state of the chip rather than to the effect of the beam itself.

In a second technique an infrared microscope was used to find the "hot spot" associated with latchup [Payne-80]. Detecting the infrared radiation given off by the latchup path, where power dissipation is high, is less invasive than stimulating hole/electron pairs with an SEM. Measurements using this technique revealed that latchup prone areas lacked well-defined substrate and P-well potentials between the P- and N-channel FET devices. This problem was corrected by adding substrate and well contacts to the critical areas.

In a third technique a laser is scanned across a wafer to identify latchup prone areas. The beam is used to both trigger and detect latchup. Locally induced photocurrents flowing through the parasitic PNPN's bypass resistors trigger latchup in circuits biased normally or subjected to an overvoltage condition. Latchup locations are indicated by bright spots in the laser-scanning microscope pictures. Measurements on both N-well and P-well CMOS reveal the most sensitive region for triggering to be at the well/substrate junction under the field oxide [Takacs-83]. The triggering mechanism is the voltage drop across the well resistance which forward biases the emitter/base junction of the vertical parasitic transistor. The current responsible for this drop consists of either photocurrent, field FET channel current, or a combination of the two. Because of drain-induced barrier lowering in the field device, the amount of photocurrent required to induce latchup decreases with increasing reverse bias on the well/substrate junction.

Computer processed EBIC measurements have been used to correlate latchup sensitivity to the spatial dopant distribution [Schwabe-84]. In a conventional deep well process, the well is counterdoped near the surface by substrate field doping, resulting in two lateral bulges in the well profile. EBIC measurements reveal both to be latchup sensitive points. In the authors' proposed shallow well process the bulges have been eliminated, and there is only one sensitive point. Plots of $V_{crit}$ (the nomenclature for switching voltage in this reference) vs. laser-induced photocurrent are used to assess latchup hardness for deep and shallow P-wells with $N^+/P^+$ spacings of 4.5 and 6.0 $\mu$m.

# Latchup Characterization

## 5.9 Characterization Review

Latchup characterization has been performed many ways, all of which have been reviewed in this chapter. For purposes of discussion, they have been grouped according to the measuring instrument used and by how many leads are presented to the measuring instrument. Usually the more leads presented, the more flexible the measurement and the more complete the data.

The latchup data most frequently displayed relates to the switching and holding points, although triggering mechanisms are also an important characterization subject, especially as they relate to the minimum $N^+/P^+$ spacing. Much of the latchup work reported in the technical literature is "relative," i.e., a measurement is made on several samples having variations in at least one processing parameter, and the results are ranked for latchup hardness. Little attempt has been made to quantitatively compare latchup data with a model for latchup except to note the dependence on bypass resistance.

The best technique for measuring the switching point places a current source on one of the parasitic emitters. Controlling emitter current controls PNPN loop gain up to the switching point and into the negative differential resistance region, clearly displaying the knee at which $dV/dI = 0$. Switching currents measured this way have been shown to be in excellent agreement with the differential latchup criterion presented in Chapter 4.

The current source forces the first transistor to turn on, and the resulting bypass current on the second transistor, if large enough, will force that transistor on. Switching current for such a configuration is described by the appropriate triode approximation. Because the bypass resistors for the two transistors are usually unequal, the measured switching current differs when the current source is placed on the other parasitic emitter.

The best technique for measuring the holding point is to first trigger the PNPN into the latched state (almost any triggering mode will do) and then to decrease the power supply until the

PNPN device switches back to the blocking state. Using a current source is good way to prevent accidentally destroying the device, and it clearly displays the knee at which $dV/dI = 0$.

# Chapter 6

## AVOIDING LATCHUP

Our explorations have taught us that avoiding latchup means operating solely in SAFE space. We have also found a definition of the switching boundary and the tools for locating the operating point in SAFE space. We now turn to CMOS design and the implementation of these concepts to prevent latchup.

This chapter focuses on design techniques for distancing the operating point from the switching boundary. They divide into two categories - layout guidelines, which transcend a particular processing technology, and process guidelines, which must be incorporated into a process technology. Most CMOS chips being designed today use at least some of these guidelines.

## 6.1 Layout Guidelines

The CMOS circuit designer interested in avoiding latchup should observe some simple layout precautions. While it is not possible to provide detailed design rules valid for all designs and all technologies, it is possible to review the basic goals of such design rules, as well as the underlying physical mechanisms.

### 6.1.1 Guard Structures

Guard structures have been used for many years to decouple one parasitic bipolar from another. There are two types, minority carrier guards and majority carrier guards, and the decoupling action differs for the two. Understanding how each works is necessary for their effective use.

Minority carrier guards are used to collect injected minority carriers before they can cause a problem. If not somehow "pre-collected," minority carriers injected into the substrate could be collected by a reverse-biased well/substrate junction and flow through the well as majority carriers. The resulting voltage drop could, if large enough, turn on a parasitic vertical bipolar in the well. An early latchup problem was traced to the input protection circuitry, which contained a $P^+$ diffusion to N-type substrate diode that was designed to forward bias when the input voltage exceeded $V_{dd}$ and to limit the input voltage swing. Unfortunately, the holes injected into the substrate caused latchup. The solution was to surround this diode with a minority carrier guard (often called a guard ring) to prevent the injected holes from reaching any P-well [Ochoa-79].

Another possible injection location is at the output of buffered driver circuits, particularly when they are used as off-chip drivers. Comparison of four stage buffers, one with a minority carrier guard ring and one without, clearly demonstrated the guard ring's efficacy [Huang-82]. Each was stressed several ways. With the input voltage at 5 V, trigger current was sourced into the output, forcing on the VPNP; with the input voltage at ground, trigger current was sourced out of the output, forcing on the LNPN. For these triggering modes the unguarded buffer exhibited latchup for trigger currents in the range of 5 mA to 10 mA, but in both cases the guard rings completely prevented latchup. The buffers also were subjected to a supply overvoltage stress. Both guarded and unguarded buffers latched, but trigger and holding currents were significantly higher when guard rings were used.

A minority carrier guard ring can be a reverse-biased source/drain diffusion or an additional well diffusion (without the opposite type source/drain diffusion so that the guard ring itself does not contain vertical parasitic transistors). Well guard rings are more effective than source/drain diffusions because they reach deeper into the substrate. In addition, minority carrier guards have been found to be substantially more effective in epi-CMOS than in regular bulk CMOS because the built-in electric field resulting from the substrate out-diffusion profile deters minority carriers from diffusing under the guard ring [Troutman-83c]. Measurements have shown that a few per hundred of the electrons injected into a $P^-$ substrate can escape from an N-well guard completely surrounding the parasitic emitter, while the escape probability for the same structure fabricated in a $P^-$ epitaxial layer on a $P^+$ substrate is reduced to a few per million.

Since carriers injected by a parasitic emitter in the well travel primarily downward to the collector, minority carrier guards at the surface have little effect on the vertical transistor. In their role as pre-collectors they can, however, substantially reduce the current gain of lateral parasitic bipolars, both in the well and in the substrate. Since the lateral bipolar in the substrate is the one comprising the most common PNPN structure (along with the vertical bipolar in the well), minority carrier guards are more frequently placed in the substrate than in the well.

To be effective a minority carrier guard ring need not completely enclose the parasitic emitter in the substrate as long as it effectively prevents any well from collecting injected charge. In many cases, however, wells are found on all sides of a potential emitter so that the guard must enclose it to prevent all possible latchup paths.

To summarize, the basic notions of the minority carrier guard ring are shown in Figure 6-1. In (a) an $N^+$ diffusion guard is placed between a parasitic emitter and the N-well. It collects electrons flowing to a depth approximately equal to the extension of its depletion region. The remaining electrons reach the N-well, and in flowing to the N-well contact create an ohmic drop that can forward bias the parasitic $P^+$ emitter. The deeper N-well guard shown in (b) is more efficient, but it requires more area than the $N^+$ diffusion. On epi-CMOS the N-well guard ring virtually eliminates electron current flow to the guarded well (c). Since guarding is more effective before the minority carriers have had a chance to spread, a minority carrier guard ring is frequently used to enclose a potential injection source.

By contrast the majority carrier guard ring decouples by minimizing the voltage drop created by majority carrier currents. As such it consist of the same type diffusion as the background into which it is placed, an $N^+$ diffusion in an N-well or a $P^+$ diffusion in a P-type substrate, for example. In essence, the majority carrier guard ring locally reduces sheet resistance.

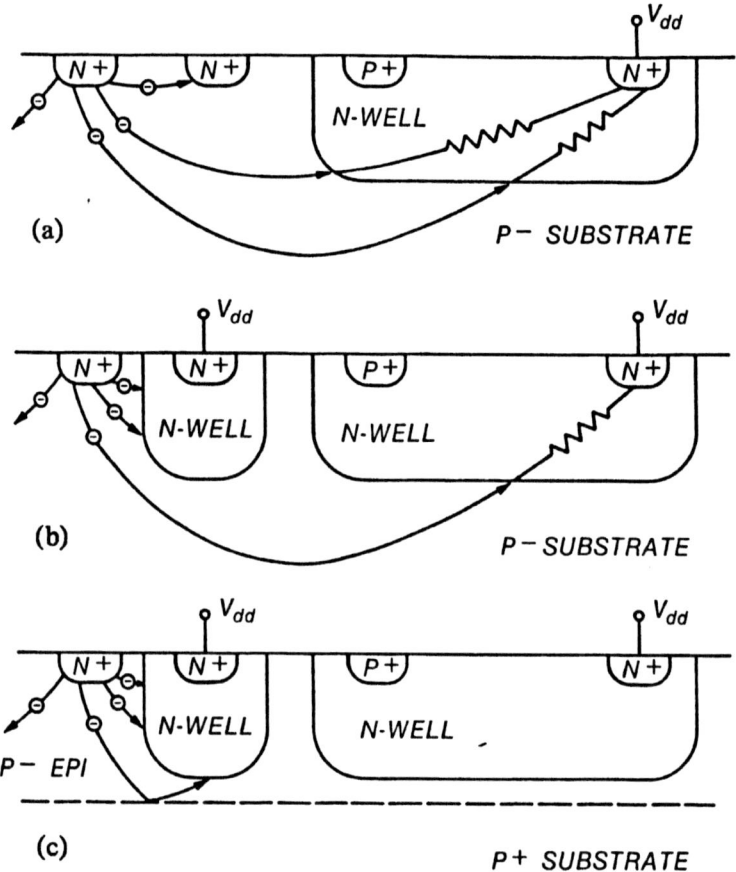

Figure 6-1. Minority carrier guard in substrate. (a) $N^+$ diffusion guard. (b) N-well guard. (c) N-well in epi-CMOS.

Majority carrier guards have been used for I/O circuits, which typically experience transient overshoot. To fully control latchup in a 3.5 μm P-well CMOS, for example, the N-channel I/O devices were surrounded with a grounded $P^+$ diffusion in the P-well, and the P-channel I/O devices were surrounded with an $N^+$ diffusion in the substrate [Payne-80]. In addition, local potentials were fixed by connecting the $P^+$ and $N^+$ guard rings to ground and power supply, respectively.

# Avoiding Latchup

Tests on a simple PNPN structure also have demonstrated the efficacy of majority carrier guards [Rung-83]. In what is termed the "weak" structure, shown in Figure 6-2(a), the four diffusions are arranged so that the $P^+$ ohmic contact to the P-well and the $N^+$ ohmic contact to the N-type substrate are farthest from one another, putting the two emitters closest to one another. By placing the ohmic contacts closest, and the parasitic emitters farthest (the so-called "strong" structure shown in Figure 6-2(b)), the diffusions forming the ohmic contacts behave as majority carrier guard rings. Holding current for the strong structure exceeded that for the weak by more than an order of magnitude, and its holding voltage exceeded 5 volts.

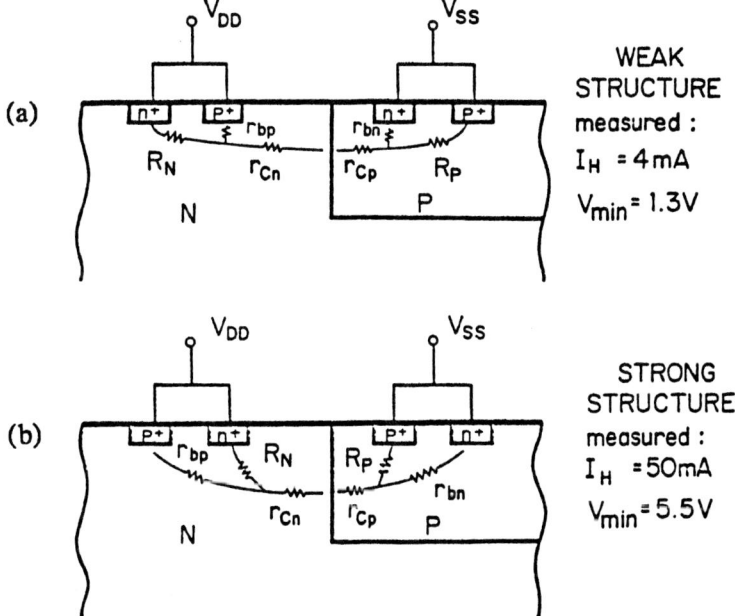

Figure 6-2. P-well PNPN structure. (a) Weak layout (more susceptible to latchup). (b) Strong layout. From [Rung-83]. © 1983 IEEE. Reprinted with permission.

The basic notions of a majority carrier guard ring in the well are illustrated in Figure 6-3. In (a) the $N^+$ diffusion forming the ohmic contact to the well is used to reduce well resistance $R_w$ for all parasitic $P^+$ emitters. Some current does flow laterally under the parasitic emitter, however. In (b) the $N^+$ diffusion is used to steer current away from the parasitic emitter, reducing the ohmic drop by reducing the LNPN collector resistance still more. In epi-CMOS this steering action is greatly enhanced (c).

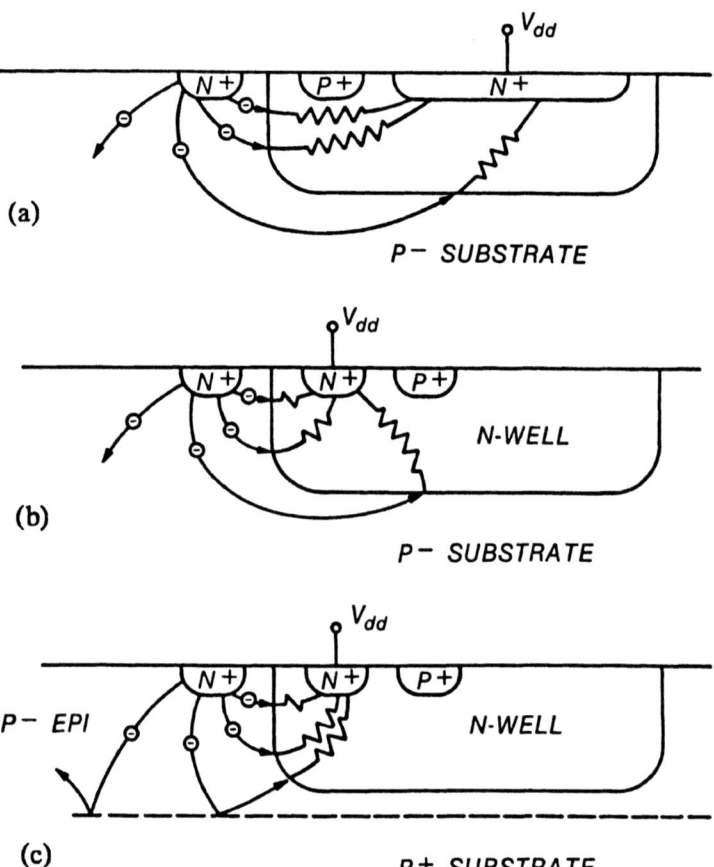

Figure 6-3. Majority carrier guard in well. (a) $N^+$ diffusion guard to reduce N-well sheet resistance. (b) $N^+$ diffusion guard to steer current away from VPNP emitter. (c) $N^+$ diffusion guard steering on epi-CMOS.

Figure 6-4 illustrates the majority carrier guard in the substrate. In (a) the $P^+$ diffusion is used to reduce substrate resistance $R_s$ for all possible $N^+$ emitters. Much of the current does flow laterally under the diffusion, however. In (b) the $P^+$

## Avoiding Latchup

diffusion is used to steer current away from the $N^+$ emitter, reducing the forward bias still more for a given current. It is especially effective against lateral base current in the well. For epi-CMOS the substrate contact ring is preferable to a majority carrier guard when the internal emitter injects mainly vertically because the highly doped substrate already has very low sheet resistance, and the distant substrate contact shifts any voltage drop away from possible emitters (c). In addition, the area of the substrate contact ring can be made much larger than that of a local substrate contact.

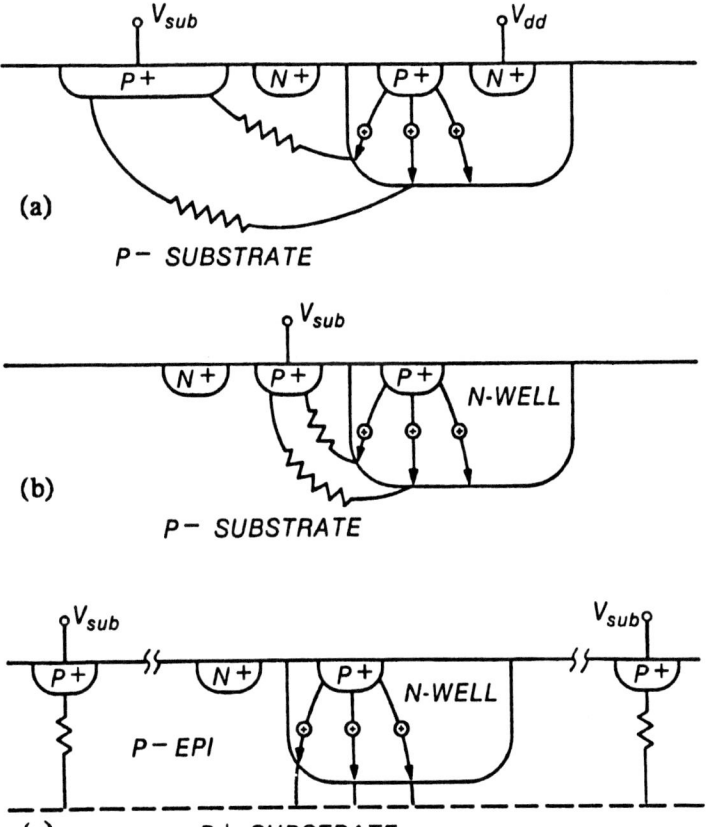

Figure 6-4. Majority carrier guard in substrate. (a) $P^+$ diffusion guard to reduce substrate sheet resistance. (b) $N^+$ diffusion guard to steer current away from LNPN emitter. (c) Contact ring is preferable to $P^+$ diffusion guard in epi-CMOS.

Similar structures have been used for transient triggering studies. Figure 5-26 on page 159 compares the forward base/emitter voltage required to trigger latchup with pulse width for different structures [Hamdy-83]. As is clearly evident, the structure using an $N^+$ guard ring in the N-well is more immune than the unguarded (weak) structure. Also including the $P^+$ guard ring in the substrate further hardens the structure.

### 6.1.2 Multiple Well Contacts

Since majority carrier currents flowing through the well are accompanied by voltage drops that can cause the vertical parasitic transistor to turn on, CMOS circuit design should limit the resistance along these ohmic paths. This can be accomplished by providing multiple contacts to the well using aluminum lines to minimize voltage variations within the well. The exact placement of well contacts depends on the well's sheet resistance and on the maximum currents expected. For a well sheet resistance of 2 K$\Omega$/□, for example, and a maximum allowable current of 50 $\mu$A, well contacts would have to be placed every 5 squares to restrict the total voltage drop between contacts to 0.5 volts. In many layouts it may be difficult to route all the wires needed for multiple well contacts, and a majority carrier guard ring in the well can be used instead. A source/drain diffusion by itself provides a sheet resistance at least two to three orders of magnitude lower than the well's, and its sheet resistance can be lowered still more by frequent contacting with a metal line.

### 6.1.3 Substrate Contact Ring

Although the highly doped substrate in an epi-CMOS technology is an effective ground plane when a backside wafer contact is used, one additional design item is required to achieve its full benefit with a topside substrate contact. A square topside substrate contact with an edge dimension comparable to the epitaxial layer thickness can add tens of kilohms of series resistance to the substrate path. Consequently, topside substrate contacts should have large areas. One means of satisfying this requirement and providing maximum dispersal of substrate current is to place a substrate contact ring around the edge of a chip. Such a ring can also provide current collection for the protection devices on the I/O pads. In conjunction with epi-CMOS a substrate contact ring can reduce the lateral bypass resistance below 1 Ω.

### 6.1.4 Butted Source Contacts

Metal connecting contiguous $N^+$ and $P^+$ diffusions across their metallurgical junction is said to form a butted contact. Such a contact can be used to minimize the emitter/base bypass resistance of a parasitic bipolar. Butted contacts have been used to tie down the N-type substrate and P-well potentials in the vicinity of parasitic $P^+$ and $N^+$ emitters [Payne-80]. In addition to minimizing bypass resistance, contiguous $N^+$ and $P^+$ diffusions in the substrate degrade injection into the substrate and spoil the lateral transistor alpha, especially at lower forward bias [Troutman-84].

Adding butted contacts to both sources in a normal CMOS inverter configuration raises holding current and voltage. Although an improvement, the ratio of holding current (or voltage) for butted contacts to that for separate, close contacts decreases with decreasing $N^+/P^+$ spacing. On a 12 μm epitaxial layer, for example, the ratio for holding currents decreases from 3X at a spacing of 20 μm to 1.5X at 7 μm while the corresponding ratio for holding voltage decreases from 2X to 1.4X [Hu-84]. However, the butted contact advantage can be improved by thinning the epitaxial layer. At the 7 μm $N^+/P^+$ spacing, holding voltage increases from 1.7 to 4.9 volts as the final epitaxial flat zone is reduced from 12 to 3.5 μm. Reversing the two diffusions in the butted contacts, so the $P^+$ diffusion in the substrate and the $N^+$ diffusion in the N-well are the closest pair, improves latchup immunity still more since these diffusions then also serve as majority carrier guard rings. Holding current for the 12 μm epitaxial layer is then raised by nearly two orders of magnitude over the normal configuration. Thus, majority carrier guards prevent latchup more effectively than do butted contacts.

The effect of a thin epitaxial layer over a ground plane is also achieved by implanting a highly doped layer below the well [Terrill-84b]. In this work the buried layer raised critical and holding currents for a standard PNPN structure from one to two orders of magnitude above that for a bulk wafer. A butted substrate contact, with its diffusions reversed (so that the substrate $P^+$ diffusion is closer to the N-well) and separated from the nearest well diffusion by 25 μm, has been used in conjunction with such a buried $P^+$ layer in N-well CMOS to raise the holding voltage above 10 volts. Holding voltage for the same structure on a regular bulk wafer was 1.2 V.

If used in combination with a thin epitaxial layer or with a majority carrier guard ring, butted contacts are helpful in preventing latchup. However, they can be used only on the source of an FET and then only if the FET is to be operated strictly in the grounded source mode. This restriction rules out its use in many logic circuits.

## 6.2 Process Techniques for Bipolar Spoiling

Process techniques for controlling latchup divide into two strategies - bipolar spoiling and bipolar decoupling. In the first, processing steps are deliberately chosen to spoil the current gain of at least one of the parasitic bipolar transistors by reducing carrier transport across its base or injection into its base. In the second, processing techniques (alone or in combination with layout techniques) are used to decouple the bipolars so that if one turns on, its collector current cannot switch on the second.

### 6.2.1 Lifetime Reduction

One of the earliest techniques of bipolar spoiling was doping the silicon with gold to reduce the base minority carrier lifetime [Dawes-76]. Gold easily diffuses throughout the silicon substrate, even at 800°C. In CMOS technologies of the late sixties and early seventies, in which the P-well depth was approximately 9 micron, a gold concentration of 1E15 /$cm^3$ typically resulted in lifetimes less than 100 nS, reducing the vertical NPN beta to less than unity. As well depth is reduced (today 3 to 5 μm depths are common), basewidth of the vertical bipolar decreases, and the requisite gold concentration increases. At about 1E16 /$cm^3$, the level required for well depths appropriate to 2 μm CMOS, the concentration of gold is so high that serious compensation results [Ochoa-79]. The increased number of generation sites also produces a sizeable leakage current.

A second technique for reducing lifetime is neutron irradiation [Adams-79]. Minority carrier lifetime (and as a consequence, parasitic bipolar current gain) depends sensitively on the damage caused by the scattered neutrons. In this work neutron fluences of 1E14 N/$cm^2$ for energies exceeding 100 keV were required for 9 micron depths. Complicating the fluence choice is the fact that anneals performed at only a few hundred degrees centigrade effectively remove the majority of divacancy-like recombination sites. The larger fluences necessary for shallower wells require unacceptably long exposure times and result in unacceptably high leakage current, particularly at high temperature [Ochoa-79].

A third technique is internal gettering. This is accomplished by first allowing oxygen to out-diffuse from the vicinity of the wafer surface, creating a so-called denuded zone where the minority carrier lifetime remains high. Then an anneal is performed to nucleate oxygen clusters in the remainder of the wafer. These clusters getter heavy metals during normal thermal processing, resulting in electrically active recombination sites. However, by measuring VNPN characteristics on a PNPN structure with the external $P^+$ diffusion connected to the N-type collector, the use of internal gettering was shown to be less effective in preventing latchup than using an $N^-$ epitaxial layer on an $N^+$ substrate [Sakai-81]. Moreover, any improvement seen for the intrinsically gettered $N^-$ substrate· at large $P^+$/P-well spacing vanished for spacings below 108 μm. This is to be expected as the lateral base width is decreased since a greater portion of the lateral transistor current remains near the surface and never encounters the recombination sites. Measurements and Monte Carlo simulations of lateral NPN beta on an intrinsically gettered wafer are in agreement and, compared to a fully denuded wafer, show a reduction ranging from 3X to 1.5X as the lateral basewidth decreases from 100 to 8 μm [Anagnostopoulos-84].

### 6.2.2 Retarding Base Field

Another method for spoiling the gain of the vertical parasitic bipolar is to build into the base a retarding electric field to impede base transport. This can be accomplished by using a retrograde well, i.e., one whose doping increases in going downward from emitter to collector.

One of the earliest proposals for a retrograde well was to use a buried $P^+$ layer under the P-well [Estreich-78]. This was accomplished by a masked boron implant prior to growth of the $N^-$ epitaxial layer. After growth, boron was again implanted to form the P-well. VNPN current gain was reduced by two orders of magnitude by the built-in electric field and by the increased Gummel number of the out-diffusing buried layer. In the presence of the built-in field, base transport limits bipolar gain, not injection efficiency.

Further studies of epitaxial buried layer CMOS have shown that for the retarding field to be effective, the product of the average electric field and width of retarding field region should exceed 100 mV [Estreich-80]. In addition, reduction of the downward transport raises lateral transport to the well edge. As noted in Chapter 4, this lateral transport can produce a larger increase of local surface potential than the vertical transport. Vertical transport is easier to deal with than lateral, especially in epi-CMOS. This lateral parasitic transistor within the well can be suppressed by including a minority carrier guard ring between the parasitic emitter and the well edge.

In other samples holding current for an inverter structure having an $N^+/P^+$ spacing of 3.5 $\mu$m increased from 2 mA for a conventional P-well CMOS to 14 mA for a heavily doped buried P-well structure [Manoliu-83]. Supply overvoltage was used to trigger latchup, and a punchthrough voltage exceeding 9 V was achieved for an $N^+/P^+$ mask spacing of 3.5 $\mu$m.

The retrograde profile can also be formed by a high energy boron implant to place the peak of the boron profile 1 $\mu$m or more below the surface [Rung-81 and Combs-82]. This implant is done after field oxidation, and only a brief anneal is performed, instead of the customary long P-well drive-in. The resulting profile is retrograde both vertically and laterally because screening by the thick field oxide moves the implanted peak closer to the surface at the field oxide periphery. Even with the retarding field associated with the retrograde profile, the VNPN gain still ranged from 30 to 50. Punchthrough and field FET current flow from the P-well to an external $P^+$ diffusion are cited as limits to reduced $N^+/P^+$ spacing.

This same technique has been used in a 1 µm N-well epi-CMOS technology [Taur-84]. A high energy (700keV) phosphorous implant places the peak of the well profile somewhat less than 1 µm below the surface. Retrograding also permits a thinner epitaxial layer because there is far less compensation of the well by the out-diffusing substrate dopant.

In another technique that produces a retrograde profile without growing an epitaxial layer, boron is implanted, driven in, and then phophorous is implanted to counterdope the surface region to a net acceptor concentration of $5E15/cm^3$ [Hashimoto-82]. Although the VNPN common emitter gain was substantially reduced from conventional P-well values, it still exceeded 100. Reported holding current was approximately 20% higher for the counter doped well. Junction depth of the compensated P-well was approximately 5 µm, and sheet resistance was less than 4 KΩ, which was half that of the conventional P-well of the same depth.

Simulations have shown that for non-epi-CMOS, switching current is typically a few tenths of a mA and little affected by the vertical transistor's common emitter current gain [Takacs-83]. Although the retrograde profile does reduce $\beta$, it often does not reduce it enough to significantly increase switching or holding current. As such, retrograded well profiles, by themselves, frequently do not provide much latchup immunity.

In summary, the retrograde well's benefit stems from the vertical retarding field developed in the base of the vertical transistor, which reduces its gain. However, one must insure adequate removal of injected holes from the well so that a parasitic PNPN is not formed involving a lateral transistor in the well, whose gain is enhanced by the retarding field. For the same reason the vertical retarding field enhances the effectiveness of minority carrier guard rings within the well, if they are used.

### 6.2.3 Schottky Barrier Source/Drain

Still another technique of bipolar spoiling is the use of Schottky barrier source/drains. Compared to diffused source/drains (especially arsenic), the emitter injection efficiency is greatly reduced. Because it is easier to form Schottky diodes to N-type silicon, it is the P-channel device in CMOS that utilizes Schottky barrier source/drains.

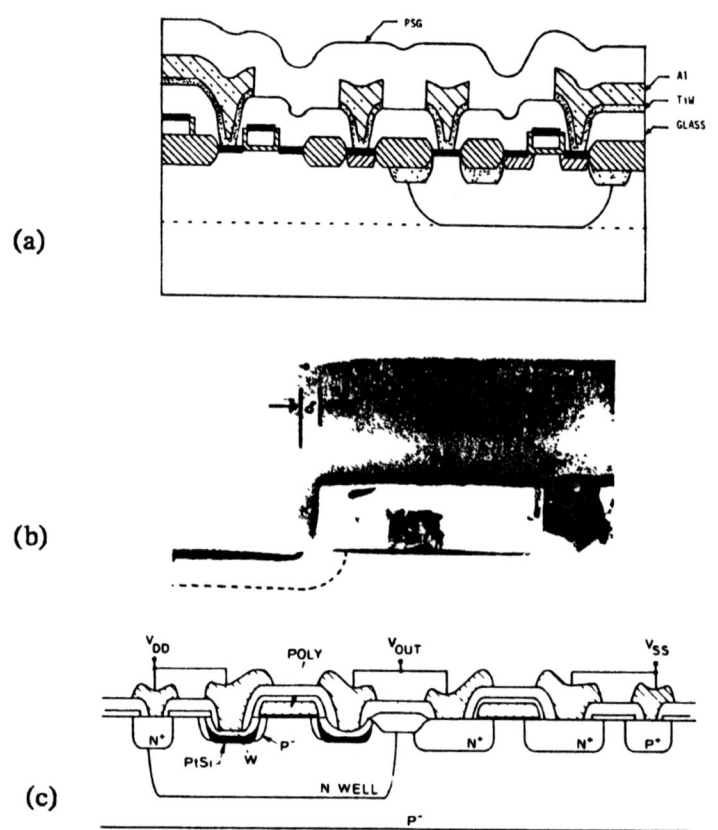

Figure 6-5. Schottky barrier source/drain fabrication. (a) Conventional Schottky barrier CMOS process. From [Sugino-82]. © 1982 IEEE. (b) TEM cross section of lightly doped Schottky MOSFET. From [Koeneke-82]. © 1982 IEEE. (c) Trenched Schottky barrier N-well process. From [Sangiorgi-84]. © 1984 IEEE. All reprinted with permission.

# Avoiding Latchup

In one implementation of the Schottky barrier technique, shown in Figure 6-5(a), platinum silicide is formed in all contacts to the silicon substrate of a P-well CMOS technology [Sugino-82,83]. The platinum silicide yields low source/drain resistance since the Schottky barrier is a metal, and junction depth can be made very shallow to combat short channel effects. When measured as a function of emitter current, beta product for a lateral PNP having a Schottky barrier contact as the emitter peaked at approximately 1E-2. The regular $P^+$ emitter samples exhibited a beta product exceeding unity over a wide range of emitter current.

Unfortunately the Schottky contact to the N-type substrate seriously degrades the FET characteristics. This degradation is a consequence of the gap remaining between the Schottky source and the edge of the gate electrode. In this gap the inversion layer must be extended toward the source by the gate's fringe field, which is much weaker than the field directly under the gate. A given gate voltage causes less current to flow in the Schottky barrier case than in a conventional, diffused source/drain PMOSFET. In addition, the Schottky barrier has higher reverse bias leakage current.

A second implementation [Koeneke-82], using PtSi in a N-well CMOS technology, has shown that a lightly doped source/drain diffusion completely surrounding the Schottky restores current drive and improves junction leakage - but at the expense of deeper junctions (contributing to short channel effects) and better minority carrier injection efficiency (resulting in higher beta for the VPNP). Its cross section is shown in Figure 6-5(b). A fundamental tradeoff exists for the dose of the lightly doped source/drain. If it is too low, the FET drive is insufficient, while if it is too high, the emitter efficiency approaches that of the diffusion rather than that of the Schottky contact. The minimum tolerable dose can probably be reduced by using P-type poly silicon gates since the accumulated layer of charge at the interface shunts the weakly doped diffusion between the Schottky contact and the gate-overlapped channel region.

A third implementation also uses PtSi contacts in N-well CMOS, but here the Schottky source/drains are trenched (recessed) into the silicon [Sangiori-84 and Swirhun-84]. After the shallow boron implant, the silicon is lightly etched in the P-channel source/drains. Platinum is then deposited, and Schottky barriers are formed at the bottom of the P-channel source/drains while the boron implant provides a self-aligned diffused guard ring laterally surrounding the Schottky diode. The finished structure is illustrated in Figure 6-5(c). This guard ring provides good ohmic contact to the inversion layer under the gate and reduces the Schottky diode's perimeter leakage. Because most injection into the base occurs in a downward direction, the Schottky barrier junction at the bottom seriously degrades the vertical PNP beta, providing good latchup immunity. The guard-ringed Schottky emitter has less minority carrier injection at its perimeter than the unguarded Schottky does at low forward bias. As emitter bias increases enough to turn on the diffused guard ring, the higher injection efficiency of the guard ring dominates, but overall emitter efficiency is still 2 to 3 orders of magnitude below a $P^+$ diffused emitter.

In this latter work, latchup is characterized on a curve tracer by displaying VPNP $I_c$ vs. $V_{ce}$ curves with the parasitic LNPN emitter shorted to the VPNP collector. Trigger current is defined as the minimum N-well (base) current required to cause "snapback" on the curve tracer for $V_{ce}$ less than 5 volts. Trenched Schottky barrier devices do not latch up until the injected current exceeds 100 mA; for the baseline process (boron implanted source/drains without Schottky barriers) only 30 $\mu$A was required. Trigger current for the trenched Schottky barrier CMOS decreases slightly with increasing boron dose.

P-channel FET characteristics for the trenched Schottky barriers are superior to those for regular Schottky barrier source/drains and approach those for conventional implanted $P^+$ source/drains. However, trenched Schottky transconductance is strongly dependent on the $P^+$ lateral doping profile, and its leakage current is somewhat higher than conventional $P^+$ source/drains and increases as the boron guard ring dose is reduced.

## 6.3 Process Techniques for Bipolar Decoupling

At present a wide variety of CMOS technologies exists with different vertical doping profiles and even different well types. This section focuses on key processing techniques and describes in general terms how each reduces latchup susceptibility. In addition, CMOS technologies are moving to micron and submicron dimensions, which will necessitate process changes. Understanding how these changes affect latchup is crucial for the process and device designers.

### 6.3.1 Epi-CMOS

The large conductance provided by a highly doped substrate is very effective in shunting the lateral transistor. If doped sufficiently highly, the substrate behaves as a ground plane, which minimizes ground bounce problems when many circuits on a chip switch simultaneously. The high/low junction formed by the lightly doped epi and the highly doped substrate also has beneficial properties. Not only does its built-in field route majority carriers into the substrate, but it also reflects injected minority carriers back into the epi, making guard rings more efficient.

The advantages of epitaxial layer CMOS have been frequently noted. In a paper discussing latchup in the then standard 9 $\mu$m P-well CMOS technology, the authors suggested using an epitaxial $N^-$ layer over an $N^+$ buried layer as the starting material for CMOS technologies having shallower P-wells [Gregory-73]. They noted the buried layer would reduce lateral voltage drops when photocurrent flows but incorrectly expected the buried layer by itself to also reduce current gain for the parasitic LPNP.

A somewhat different approach was introduced in the epitaxial buried layer P-well CMOS [Estreich-78]. The emphasis here was on a buried $P^+$ layer to provide a retrograde well profile although the advantage of a low resistivity substrate reducing lateral substrate resistance was also noted. The built-in field and increased Gummel number associated with the retrograde well profile reduced the VNPN gain by two orders of magnitude. The epitaxial buried-layer CMOS was found to be much less susceptible to latchup than regular bulk P-well CMOS [Ochoa-78]. It provided latchup immunity under normal bias conditions but not when base/emitter junctions are deliberately forward biased, in which case guard rings are also necessary.

Next, an $N^-$ epitaxial layer on an $N^+$ substrate without any buried layer was proposed as the starting material [Schroeder-80]. The $N^-$ on $N^+$ substrate does not lead to increased leakage, as does lifetime spoiling, and it was seen as more applicable to reduced geometry designs. The major design goal cited was to increase holding current enough so that parasitic bipolar gains fall off (because of high level injection) to the point of not sustaining regeneration. Ionizing radiation and supply overvoltage both triggered latchup in an 8-bit ALU fabricated in non-epi-CMOS while the same circuit in epi-CMOS remained latchup free.

The $N^-$ epi on $N^+$ substrate eliminated latchup on chips designed in a 7.5 µm P-well technology [Payne-80]. However, field failures of packaged chips using a 5 µm technology pointed out the need for butted contacts in I/O circuitry to better secure substrate and P-well potentials in the vicinity of possible $P^+$ and $N^+$ emitters. These failures also uncovered the need for a package connection to the substrate (rather than to the epi alone) in their custom dual inline package. In addition, they reported the need for majority carrier guard rings around I/O circuitry in their 3.5 µm CMOS.

The above considerations also apply to $P^-$ epitaxial layers on $P^+$ substrates for N-well CMOS. Latchup comparisons between epi and non-epi N-well CMOS were made for a standard 4-stage buffer, a $N^+/P^+$ shunted buffer (the wide N- and P-channel inverter devices are orthogonal to one another), and a guard-banded buffer [Huang-82]. Triggering by current injection into the substrate (the lateral base) did not cause latchup in the epi case for any of the buffers or for the guard-banded buffer in the non-epi case. Triggering using a curve tracer to provide an overvoltage always caused latchup, but both trigger and holding currents increased using the guard band and using a lightly doped epitaxial layer on a highly doped substrate.

The $N^-$ on $N^+$ substrate has also been used as the starting material for twin well CMOS [Parillo-80, Wieder-81, and Schwabe-83]. In such a process, initial doping of the epi layer is much lighter to allow for more optimum N- and P-channel FET's. The shunting ability of the highly doped substrate is still operative in controlling latchup.

Although it has been long recognized that the highly doped substrate in epi-CMOS effectively shunts the lateral parasitic bipolar, boosting switching current, epi-CMOS provides a second important benefit. Using a structure with two concentric N-well collectors that both completely surround an $N^+$ emitter, the escape probability of an injected electron from an N-well guard ring has been shown to be at least three to four orders of magnitude lower for epi-CMOS compared to non-epi [Troutman-83].

In another comparison of epi vs. non-epi-CMOS, the switching current for epi-CMOS was found to be two orders of magnitude larger and essentially surface controlled, the latter a consequence of field FET action and the influence on lateral bipolar gain [Takacs-84]. Latchup sensitivity decreased with increasing distance from the P-well edge to both external $P^+$ diffusions and internal $N^+$ diffusions, with decreasing epitaxial layer thickness, and with increasing substrate doping.

Similar results have also been demonstrated for N-well [Hu-84b]. Epi-CMOS raises holding voltage and current over the bulk case, and for a final epitaxial flat zone thickness of 12 $\mu$m, holding current and voltage each increased with larger $N^+/P^+$ spacing. Reducing the final epitaxial flat zone thickness from 12 $\mu$m to 3.5 $\mu$m substantially raised holding voltage (to greater than 5 V for $N^+/P^+$ spacings exceeding 7 $\mu$m).

In a study of latchup dependence on shallow P-well processing parameters, supply overvoltage triggering and vertical base current triggering measurements were made for 0.8 and 1.5 $\mu$m P-wells on $N^-$ epitaxial layer thickness ranging from 1 to 5 $\mu$m [Lewis-84]. The higher beta of the shallower well yielded lower trigger currents, and overvoltage induced latchup occurs at lower voltage because of collector/emitter punchthrough. Holding current and voltage are nearly independent of well depth. Reducing the lateral spacing of vertical emitter to well edge increases latchup susceptibility since it increases the influence of surface currents in the substrate. In this work thinner epi was shown to reduce trigger (but not holding) current, probably due to an increased Early effect.

Epi-CMOS enjoys additional benefits and adds less than 5% to the final wafer cost [Kokkonen-84]. It improves the surface lifetime for dynamic RAM memory chips and reduces their soft error rate when the storage node is placed within the well. (The epi-CMOS reduces the usual electric field collapse at a reverse biased junction accompanying an alpha particle track.) It also reduces the DC resistances for EPROM and logic chips.

In a variation of the epi-CMOS technique, a high energy (4 Mev), moderately high dose blanket boron implant is used to create a buried $P^+$ layer under the N-well to lower substrate sheet resistance [Terrill-84b]. Both critical current and holding current increase with dose, and at a dose of $1E14/cm^2$ the critical current is 90X (and holding current 30X) that for a bulk wafer when the $N^+/P^+$ spacing is 10 μm.

### 6.3.2 Retrograde Well for Lower Sheet Resistance

Just as the highly doped substrate can be used to reduce substrate resistance and shunt the lateral bipolar, so a retrograde well can be used to reduce well resistance and shunt the vertical device. This translates directly to fewer well contacts when a given layout is trying to guarantee a limit on well resistance between well contacts or from a well contact to an internal source/drain diffusion. Because of practical limitations on the peak concentration, sheet resistance for a retrograde well still exceeds that for a conventional well containing a majority carrier guard diffusion. Because the well is inherently much shallower than the substrate thickness, the benefits of a retrograde well are not expected to be as great as epi-CMOS. The retrograde well could also prove to be the key to successfully harnessing the parasitic vertical bipolar.

There are several possible ways of fabricating a retrograde well - buried layer epitaxy, high energy double implantation, and

# Avoiding Latchup

counter doping implantation. Each was discussed in section 6.2.2, which addressed the bipolar spoiling effect of a retrograde well resulting from the retarding base field.

### 6.3.3 Substrate and Well Bias

Electrical isolation is also provided by biasing the substrate below $V_{ss}$ in N-well CMOS. All potential $N^+$ emitters in the substrate are then reverse biased. The ohmic drop in the substrate required to turn on the LNPN then increases from approximately 0.5 V to 0.5 plus the amount of reverse bias. Similarly, biasing the N-well above $V_{dd}$ reverse biases all potential $P^+$ emitters and makes it more difficult for an ohmic drop in the well to turn on the VPNP.

An on-chip voltage generator is frequently used to eliminate the need for a separate power supply and at least one extra I/O pin. A substrate generator in N-well CMOS uses charge pumping circuitry to accumulate charge on the capacitance coupling the substrate to AC ground. This substrate capacitance consists of all well/substrate junctions, all $N^+$ source and drain junctions, and any stray capacitance from the substrate to ground and power supply leads and to the package. Likewise, a well generator builds up charge on the capacitance coupling the well to AC ground.

Figure 6-6 schematically illustrates a regulated two stage substrate generator [Piro-85]. The diode usually consists of N- or P-channel FET's, and the forward diode drop is really the FET threshold. When node Q goes to $V_{dd}$, diode D1 conducts d and capacitor C1 charges toward $V_{c1} = V_{dd} - V_{d1}$, where $V_{d1}$ is the forward diode drop across D1. When Q goes to ground, the voltage on node A is initially forced down (ideally by the voltage $V_{dd}$) so that D1 is shut off and diode D2 conducts. Capacitor C2 then charges toward $V_{c2} = (2V_{dd} - V_{d1} - V_{d2})$, where $V_{d2}$ is the forward diode drop across D2. When Q goes high again, both D1 and D3 conduct, and the substrate capacitance charges toward $V_{sub} = -(2V_{dd} - V_{d1} - V_{d2} - V_{d3})$. Many operating cycles are needed to achieve the voltages mentioned above. When the substrate voltage reaches a pre-determined level, the regulator disables the oscillator to save power.

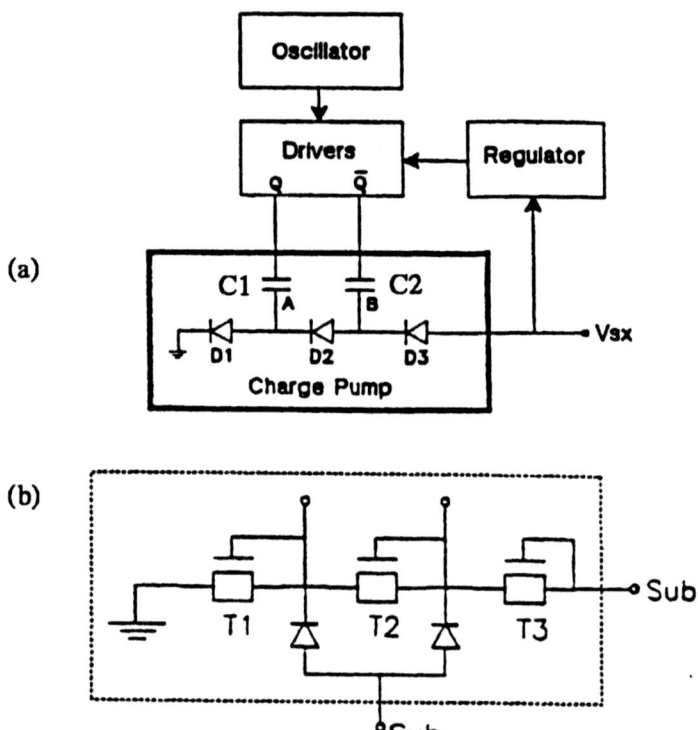

Figure 6-6. Substrate generator with 2-stage charge pump. (a) Block diagram; example shown uses a regulator to save power. (b) Schematic of N-channel implementation. From [Piro-85]. Reprinted with permission.

Figure 6-7 shows some important voltage waveforms when N-channel FET's are used for diodes as the pumping reaches the pre-determined level of -3 . The drain junctions of these FET's add two more diodes to the circuit, as indicated in Figure 6-6 (b). Note from the waveforms that node B is more negative than the substrate during part of the cycle, which causes electron injection into the substrate. When collected by a nearby N-well containing $P^+$ diffusions, these electrons form a majority carrier current that, if large enough, would turn on a VPNP. This can easily be prevented by surrounding the diodes with a minority carrier guard ring (such as an N-well without any $P^+$ diffusions and biased at $V_{dd}$) to collect the injected electrons before they cause a problem.

# Avoiding Latchup

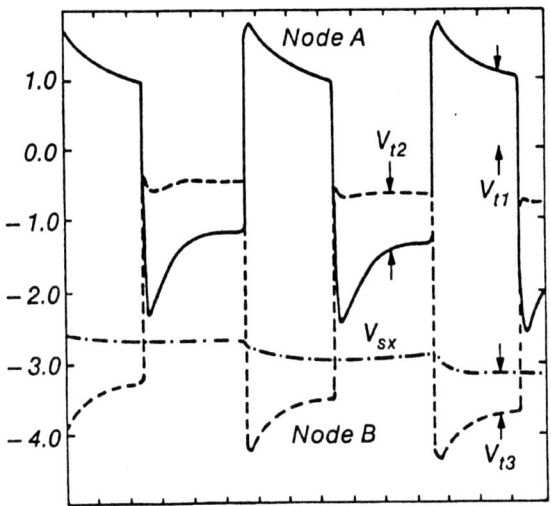

Figure 6-7. Charge pump waveforms. From [Piro-85]. © 1985 IEEE. Reprinted with permission.

Using P-channel FET's for the diodes also would avoid charge injection because all the $P^+$ diffusions are located in an N-well biased at $V_{dd}$, and their voltages are at or below ground. However, the greater back gate bias effect of P-channel devices in N-well CMOS raises the forward voltage drop and makes them less efficient than N-well diodes. There is no forward biasing problem with N-channel diodes used for generating bias on an N-well, but raising voltage too much could result in avalanche breakdown of an N-channel drain diffusion.

Substrate generators can function in the presence of only a limited amount of leakage current or noise. Multiple noise spikes from the power supply or other noise sources can be cumulative in their effect of reducing substrate bias [Hall-84]. Whereas a single spike may not cause latchup, a sequence of spikes could if they occur too rapidly for the charge pump to recover.

A worse situation occurs if a chip (or card) is inserted "hot", so that signals can appear at input pads before the substrate generator has had time to apply any bias. Without a reverse bias on its base, the parasitic lateral device is more likely to turn on.

Turn-on of the lateral bipolar constitutes at least temporary latchup. Sustained latchup can be prevented by decoupling the lateral device with a minority carrier guard ring around any lateral devices likely to turn on during hot insertion or with a majority carrier guard structure in the well containing the vertical parasitic devices likely to be switched on by the resulting lateral collector current. However, even temporary latchup could result in permanent damage, and the situation described above should be avoided by proper sequencing.

### 6.3.4 Trench Isolation

Within recent years it has become apparent that the recessed oxide (or LOCOS) FET isolation technique has reached the limits of down-scaling. One of the proposed replacements is trench isolation. Because the trench does not support current flow, it was hoped trench isolation would also prevent latchup.

Direct comparison of PNPN structures using trench isolation on both epi and non-epi substrates have shown that trench isolation by itself does not sufficiently control latchup [Rung-82b]. Trench isolation made possible the 1.2 $\mu$m $N^+/P^+$ spacing shown in Figure 6-8(a), but latchup still occurred on the non-epi sample. No latchup occurred for the epi sample as long as good contact was made to the wafer. In conjunction with a lightly doped epitaxial layer on a highly doped substrate, trench isolation promises to be an effective means of latchup control since it eliminates lateral current to and from the well, a potentially significant triggering current at small $N^+/P^+$ spacings. Unfortunately, trench isolation at minimum $N^+/P^+$ spacings in actual layouts causes additional problems. Oxide charge along the poly-silicon filled trench sidewall causes parasitic current paths when two or more $N^+$ diffusions are butted against the trench.

In oxide-filled trenches the problem is even worse. When one side of the trench is bounded by the P-type substrate and the opposite side is bounded by the N-well, as is illustrated in Figure 6-8(b), the biased (at $V_{dd}$) N-well acts as a gate and tends to invert the substrate along the trench wall [Cham-83]. This causes leakage and field FET problems. When the trench surface is inverted, LNPN current gain also is higher than for the non-trench structure, probably because the effective collector

# Avoiding Latchup

then includes the inverted trench surface. Clearly, some form of channel stopper is required for trench isolation.

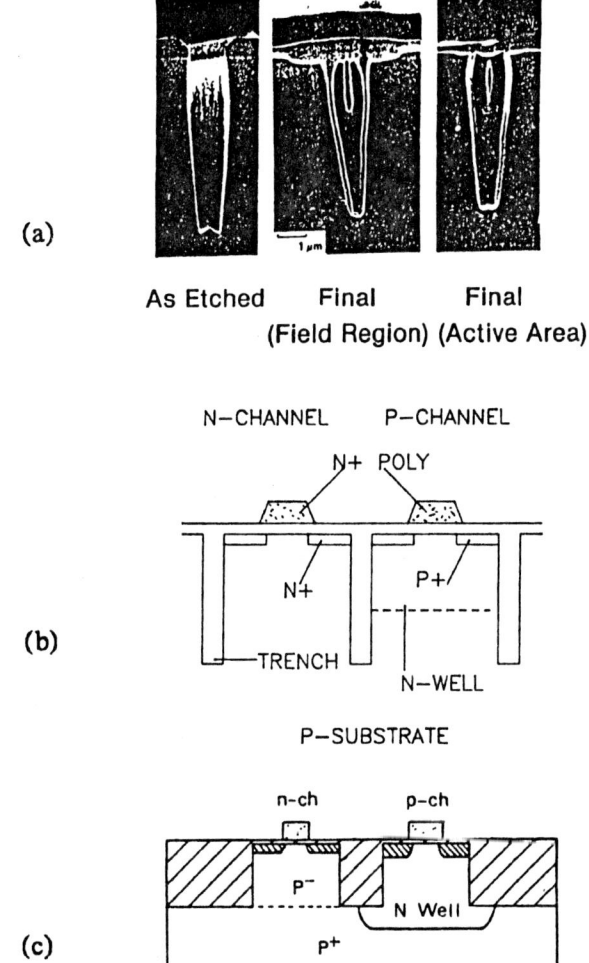

Figure 6-8. Various lateral isolation technologies. (a) SEM view of polysilicon filled trench. From [Rung-82b]. © 1982 IEEE. (b) Cross section of oxide filled trench. From [Cham-83]. © 1983 IEEE. (c) CMOS cross section using selective epitaxial growth. From [Endo-83]. © 1983 IEEE. All reprinted with permission.

If one desires the minimum $N^+/P^+$ spacing allowed by lithography for sub micron CMOS, it may be necessary to use some form of trench isolation at the well edges to insure there is no lateral transistor action from an internal source/drain diffusion to the substrate. A trench as deep as, or deeper than, the well in an epi-CMOS process might provide a lithographically limited $N^+/P^+$ spacing. However, current problems with inversion of the vertical sidewalls must be solved before diffusions can be butted against the trench walls.

Another promising lateral isolation technique is selective epitaxial growth, which has been proposed in conjunction with a $P^+$ substrate to control latchup [Endo-83]. A thick oxide (2 μm) is first grown over the entire wafer, then delineated into isolation areas by reactively sputter etching down to the silicon substrate. LPCVD poly silicon or plasma CVD amorphous silicon is deposited on the vertical sidewalls of the oxide pattern to provide a channel stopper, which, incidentally, resulted in less electrical width bias for the FET's than the conventional LOCOS process. A $P^-$ epitaxial layer is then grown selectively on the exposed silicon $P^+$ wafer, and an N-well formed by a phosphorous implant followed by a drive-in. The final result is shown in Figure 6-8(c).

## 6.4 CMOS Design Considerations

One conclusion to be drawn from the preceding discussion is that there are several ways of avoiding latchup. However, in practice not all choices are equally likely. Other considerations are at least as important as avoiding latchup, necessitating design tradeoffs. Here we summarize the fundamentals of latchup avoidance and review key design options.

For a CMOS receiver/driver circuit, an I/O overshoot can turn on the parasitic PNP, injecting holes into the N-well. The injected holes are collected by the N-well/substrate junction. On epi-CMOS the holes collected at the N-well bottom flow into the highly doped substrate. As long as a low impedance path is provided to the highly doped substrate (via a backside contact or a topside contact ring), the VPNP collector current cannot produce a sufficient potential drop to turn on a parasitic LNPN.

# Avoiding Latchup

Thus, epi-CMOS in conjunction with a low impedance substrate contact decouples the LNPN from the VPNP.

There is the possibility that the internal $P^+$ diffusion is close enough to the N-well edge that a significant portion of the injected holes are collected by the adjacent N-well sidewall. This current is not as effectively shunted by the highly doped substrate and could cause an ohmic drop sufficient to turn on the LNPN. Care must be exercised in the process design to avoid significant sidewall collection. The transmission line model can be used to ascertain the tolerable sidewall component as a function of external $N^+$ spacing once the vertical doping profile is known for the substrate.

An I/O undershoot on a receiver/driver circuit will not be a problem if a negative bias is applied to the substrate, and the magnitude of the applied bias exceeds the undershoot. However, if this condition is not met, as for grounded substrate operation, the undershoot could turn on a LNPN. There are then basically two ways of decoupling the LNPN and preventing the injected electrons from turning on a VPNP.

The first is to interpose a minority guard between the injecting $N^+$ diffusion and any N-well containing a $P^+$ diffusion. Such a guard would be either an $N^+$ diffusion or an N-well (lacking $P^+$ diffusions) reverse biased with respect to the substrate.

The second is to use a majority carrier guard within the N-well to reduce the VPNP bypass resistance. Surrounding the N-well with an internal $N^+$ diffusion provides shunting action no matter where along the N-well edge the electrons are collected. This guard ring may be simply the $N^+$ diffusion added to the N-well to make good ohmic contact. For a long guard ring multiple contacts to the $V_{dd}$ line may be needed to guarantee ohmic drops less than a few tenths of a volt.

In the first case above, injected electrons are pre-collected before they can reach an N-well containing $P^+$ diffusions, and in the second electrons arriving at such an N-well are shunted from the N-well proper by the lower impedance path offered by the internal $N^+$ diffusion. Which is to be used is a matter of geometry and application. When there are many receiver/driver circuits, one minority carrier guard ring around them all will probably cost less area than majority carrier rings on each and every N-well, especially if there are many relatively small, unconnected wells. In addition, there may be other good reasons for not wanting excess electrons in the substrate, such as disturbance of pre-charged nodes or DRAM cells. On the other hand, there may be only a few large N-wells containing $P^+$ diffusions, and employing majority carrier guard rings may cost less area. Also, the majority carrier guard ring requires no extra routing of $V_{dd}$. Of course, both types of guards can be employed where the areal cost is tolerable.

## 6.5 Latchup-free Design

We have already discussed why bipolar decoupling is preferable to bipolar spoiling for preventing latch-up. Now we want to discuss in general terms why this decoupling can yield latchup-free design. First we note the triggering modes discussed in Chapter 3 can be grouped according to whether the vertical or the lateral parasitic bipolar is the first to be turned on:

• TVP - Some external triggering mechanism turns on the vertical parasitic transistor, which can then provide the current to turn on the lateral parasitic transistor, resulting in latchup; and

• TLP - Some external triggering mechanism turns on the lateral parasitic transistor, which can then provide the current to turn on the vertical parasitic transistor, resulting in latchup.

Latchup is sustained if the second transistor continues to activate the first after the original triggering is removed. If the second transistor cannot supply sufficient current to hold the first on, both transistors switch off after the original triggering is removed, and the latchup is then said to be unsustained or temporary. Even temporary latchup can produce permanent damage or circuit malfunction since it momentarily provides a low impedance path from the power supply to ground.

Likewise, it is convenient to group the decoupling strategies according to which parasitic bipolar is turned on first:

- DVP - The vertical parasitic transistor is decoupled from the loop by not allowing its collector current to turn on the lateral parasitic transistor.

- DLP - The lateral parasitic transistor is decoupled from the loop by not allowing its collector current to turn on the vertical parasitic transistor.

Each of these decoupling strategies can be accomplished several ways.

### 6.5.1 Decouple Vertical Parasitic (DVP)

The strategy here is to prevent the vertical parasitic from turning on the lateral by designing a sufficiently small substrate bypass resistance. The lateral parasitic bipolar is then shunted so completely that collector current from the vertical parasitic cannot create enough voltage drop to turn on the lateral. Specifically, this consists of epi-CMOS, a substrate ring contact, and, in the fully optimized case, substrate out-diffusion to the bottom of the well in order to minimize the epitaxial layer flat zone thickness.

The effectiveness of this strategy can be quantified once the $\beta_{fp}$ falloff with collector current $I_{cp}$ is known. A common description of this behavior is given by (see equation 2.193 of [Getreau-78], for example),

$$\frac{\beta_{fp}^o}{\beta_{fp}} = 1 + \frac{I_{cp}}{I_{kfp}}, \qquad (6-1)$$

where $\beta_{fp}^o$ is the peak, large-signal common emitter current gain and $I_{kfp}$ is the knee current (see section 4.3.4) for the VPNP. It is easily shown that the analagous description for the small-signal current gain is

$$\frac{\beta^o_{fps}}{\beta_{fps}} = 1 + \frac{2I_{cp}}{I_{kfp}}. \qquad (6-2)$$

To continue the example, assume the bypass resistance in the well is larger than that in the substrate and use the floating-well triode approximation. Then equation (4-57) defines the LNPN base/emitter forward bias $V_{ben,s}$ required to reach the switching point. Also assume the current $I_{ci}$ (defined in section 4.7.1) to be negligible. Then, the equivalent LNPN bypass resistance is the output transfer resitance $R_{to}$, which is bounded on the high side by the output resistance $R_2$ for all values of transmission line parameters. Consequently, the worst case surface potential induced at the edge of the LNPN emitter is

$$\psi_{sno} = R_2 I_{cp}. \qquad (6-3)$$

The design strategy can now be stated simply as $\psi_{sno} < V_{ben,s}$, or in terms of the output resistance,

$$R_2 < \frac{2V_{ben,s}}{(\frac{\beta^o_{fps}}{\beta_{fps}} - 1)I_{kfp}}, \qquad (6-4)$$

which provides a design equation stating how small the output resistance need be to avoid latchup. The constants $I_{kfp}$ and $\beta^o_{fps}$ can be easily determined for a particular CMOS technology, but what value of $\beta_{fps}$ is to be used?

The condition for remaining in SAFE space with a floating N-well can be found from equations (4-45), (4-46), and (4-47) to be

$$\frac{r_{en}}{R_s} > \frac{\alpha_{fps} - (1 - \alpha_{fns})}{1 - \alpha_{fps}}. \qquad (6-5)$$

## Avoiding Latchup

As $I_{cp}$ increases above the knee current $I_{kfp}$, the small-signal current gain $\alpha_{fps}$ decreases. Once it falls to a value $\alpha_{fps} = 1 - \alpha_{fns}$, latchup is no longer possible. Thus, the proper $\beta_{fps}$ value to be used in equation (6-4) is its minimum value such that latchup is still possible, or $\beta_{fps} = 1/\beta_{fns}$. Under worst-case conditions for the LNPN, its current gain would be at the peak value, so that equation (6-4) then becomes

$$R_2 < \frac{2V_{ben,s}}{(\beta^o_{fps}\beta^o_{fns} - 1)I_{kfp}}. \qquad (6-6)$$

A typical value for the right hand side can be found from the data in Figure 5-11 on page 141. There the values $\beta^o_{fps} = 41.8$ and $\beta^o_{fns} = 7.06$. Using equation (6-2) to fit the curve of $\beta_{fps}$ vs. $I_{ep}$ yields $I_{kfp} = 1.5$ mA. Substituting the maximum alpha values into equation (4-57), as well as $I_{sn} = 7.9$fA (from section 5.4) and a bypass resistance of 10 $\Omega$, yields $V_{ben,s} = 0.6$ V. Finally, substituting the above values into equation (6-6) yields $R_2 < 2.7$ $\Omega$ as the desired quantitative design equation.

Following such a strategy in N-well CMOS guarantees that neither temporary nor sustained latch-up occurs when only the VPNP is triggered because it cannot turn on the LNPN. If, on the other hand, the LNPN is triggered and if it in turn generates enough collector current to turn on the VPNP, temporary latch-up can occur. Removal of the external triggering allows the LNPN to turn off since the VPNP cannot provide enough current to hold the LNPN on. Of course, once the LNPN turns off, so does the VPNP.

Because of its high common emitter current gain, transient turn-on of the parasitic vertical bipolar can draw large current spikes from the power supply. While this may not cause latchup if the vertical device is properly decoupled, such large spikes waste power and could cause other problems. Any circuit expected to encounter harsh noise conditions, such as off-chip drivers, should use majority carrier guard rings in the well and employ ample well contacting to prevent the parasitic vertical bipolar from turning on.

### 6.5.2 Decouple Lateral Parasitic (DLP)

The strategy for this case is to prevent the lateral parasitic from turning on the vertical. One method to accomplish this is to make the sheet resistance for the well as low as practical and use multiple well contacts, if needed, to reduce the number of squares for current flowing through the well. A majority carrier guard ring in the well helps reduce sheet resistance; so does a retrograde well profile. A second method is to place minority carrier guard rings outside the well to pre-collect any carriers injected by the parasitic lateral bipolar before they reach the well. Details of this technique can be quantified once the $\beta$ falloff for the lateral parasitic bipolar is known as a function of collector current.

Following such a strategy in N-well CMOS guarantees that neither temporary nor sustained latch-up occurs when only the LNPN is triggered because the LNPN cannot turn on the VPNP. If, on the other hand, the VPNP is triggered and if the VPNP generates enough collector current to turn on the LNPN, temporary latch-up can occur. Removal of the external triggering allows the VPNP to turn off since the LNPN cannot provide enough current to hold the VPNP on. Of course, once the VPNP turns off, so does the LNPN.

Effectively implementing both DVP and DLP strategies insures that no matter which parasitic bipolar is turned on, it cannot turn on the second, so that latchup cannot occur. Figure 6-9 summarizes the worst case latch-up consequences for each decoupling strategy alone and in combination. Note that decoupling either parasitic transistor prevents sustained latchup.

|     | No Decoupling | DVP | DLP | DVP + DLP |
| --- | --- | --- | --- | --- |
| TVP | Sustained | No Latchup | Temporary | No Latchup |
| TLP | Sustained | Temporary | No Latchup | No Latchup |

Figure 6-9. Worst case latchup consequences.

# Chapter 7

## SUMMARY

After our long and sometimes arduous journey, we should reflect a bit before concluding. Where have our explorations taken us? What have we learned? How can we use this knowledge?

## 7.1 Problem Description

CMOS has always been troubled, fatally for some chip designs, by a problem inherent to its very structure. Fabricating the complementary N- and P-channel FET's also produces parasitic PNPN devices that can exhibit a circuit-threatening, low-impedance, latched state in addition to the high-impedance blocking state. Such PNPN devices are susceptible to latchup because during normal operation various stimuli can trigger switching from the blocking state.

To aid latchup analysis and characterization, these stimuli, or triggering modes, can be grouped into three categories: (1) external initiation of the first bipolar, (2) normal bypass current initiation of both bipolars, and (3) degraded bypass current initiation of both bipolars. This grouping helps distinguish which PNPN configuration is being stimulated and which bypass resistor more effectively determines the switching current.

Latchup occurs if the loop gain for the relevant PNPN configuration exceeds unity, if both parasitic bipolars are turned on long enough for the PNPN current to reach the switching current, and if the power supply and associated circuits supply at least the switching current (to leave the blocking state) or the holding current (to reach the latched state). Latchup is sustained if the PNPN remains in the latched state after removing the stimulus, and is only temporary if it does not. Simply leaving the blocking state, even temporarily, can disrupt normal circuit operations. In a healthy CMOS technology, no parasitic PNPN device ever leaves the blocking state.

## 7.2 Modeling and Analysis

PNPN device modeling actually began several years before CMOS was invented. Theoretical studies of the hook collector transistor led to the notion of using a bypass element to make current gain more uniform. Modeling progressed with the semiconductor controlled rectifier (SCR), a PNPN structure designed as a switching device. In its early development, avalanching the blocking junction was used to switch the SCR from its high to its low impedance state, but supplying a gate current via an external base lead was soon found to be quicker and more efficient. Positive gate current switched the device from high to low impedance, negative gate current from low to high. This negative gate current, or bypass current, was used to provide a blocking state when the sum of small-signal alphas exceeded unity.

The lumped element model for the parasitic PNPN structure is a direct descendent of the SCR lumped element model. This model has been very helpful in qualitatively understanding latchup behavior, but it was soon discovered the lateral parasitic bipolar can behave differently when operating together with the vertical parasitic bipolar than when operating alone.

Although the lumped element model has been a useful analysis tool, estimating its lateral bypass resistance is complicated by two-dimensional current flow in the substrate. Whether a given substrate current can initiate latchup depends on the voltage generated in the vicinity of a potential substrate emitter. Fortunately, the substrate can be viewed as a lossy transmission line, and calculating the desired voltage (or current) changes in the substrate for a given layout reduces to a simple transmission line problem. The line's characteristic length and impedance are relatively simple functions of the vertical doping profile from the field region to the highly doped substrate. These results are used

# Summary

to define the transfer resistances $R_{to}$ and $R_{ti}$ for any layout configuration in any epi-CMOS technology.

Our understanding of when latchup occurs has evolved over the years, as have the criteria used to define it. This book shows that the goal of any latchup prevention strategy should be to restrict operation to the blocking state. A new, differential latchup criterion has been derived which states that switching from the blocking state will not occur as long as the sum of the effective small-signal alphas remains less than unity. One particularly useful implementation of this criterion, which applies where series base and emitter leg resistance are negligible, is

$$\frac{\alpha_{fns}}{1 + \frac{r_{en}}{R_s}} + \frac{\alpha_{fps}}{1 + \frac{r_{ep}}{R_w}} < 1. \qquad (7-1)$$

Where necessary, high level injection effects are easily included in the small-signal emitter resistance, as are emitter and base leg resistances. The general form of this criterion is obtained by substituting equations (4-45) and (4-46) into (4-47).

The differential latchup criterion also introduces the concept of SAFE space, which is a pictorial representation of the blocking state. Operating trajectories through SAFE space illustrate how triggering occurs. For a PNPN triode configuration the trajectory is a straight line perpendicular to one of the axes. The locus of operating points begins at the intercept on the axis and ends at the switching boundary. For a PNPN tetrode the trajectory is a curved line bounded by the trajectory for the triode that results when the larger bypass resistor is removed. Now the locus begins at the origin and ends on the switching boundary. The triode configuration is an excellent approximation for the tetrode when one bypass resistor exceeds the other by 2X. The appropriate triode replacement is obtained by removing the larger bypass resistor with the result that one of the parasitic bipolars is not operating in the bypassed mode.

The differential latchup criterion also leads to a precise expression for switching current, which is the total current through the PNPN device when the operating point is on the switching boundary of SAFE space. Switching current is uniquely determined by, and easily related to, the relevant PNPN configuration. The triode approximation is very useful for simplifying the expression for switching current and obtaining a simple and accurate design equation. In equation (7-1) the larger bypass resistance is taken to be infinity as a starting point in its derivation.

Another distinctive and important point on the PNPN current/voltage characteristic is the holding point. However, the equation commonly cited in the literature as "holding current" neglects reverse transistor action. As such, it is a valid description of device current when both parasitic transistors are turned on but only for PNPN device current less than, or equal to, the turn-off current $I_{to}$. Exactly which point on the current/voltage characteristic it represents depends on the base/emitter voltage substituted into the equation. For device current $I > I_{to}$, the reverse current must be taken into account, even if there is no reverse transistor action.

Modeling is somewhat easier for the switching point than for the holding point. The condition of the parasitic bipolars is much better determined since each is either off or in the forward active mode, making modeling more definite. In the latched state at least one bipolar is in saturation, which can drastically alter current flow. Whether both are saturated depends on the $R_{s2}$ and $R_{w2}$ values for interconnection resistance. In addition, high-level injection effects are more likely to be important in the latched state, and conductivity modulation can lower effective resistances. Finally, switching current is independent of the $R_{s2}$ and $R_{w2}$ values, which influence both holding current and voltage.

Type 2 triggering was used to illustrate latchup in a lumped element model. In agreement with the differential latchup criterion, latchup occurred when the effective small-signal alphas reached unity. The corresponding sum of static alphas at the switching point was approximately 0.7. Simulated holding current was 1.47 times the turn-off current and 2.51 times the switching current.

The actual triggering waveform is important for understanding whether latchup occurs since the dynamic operating point must reach the switching boundary of SAFE space. Because of finite base transit times and possible RC time constant limits on applied pulses, any excitation must be applied for a critical length of time, which is the time to bring the PNPN device current up to the switching current. Excitation applied for a shorter time, regardless of magnitude, does not cause latchup. When applied for times much longer than the critical time, the required bypass current reduces asymptotically to the value calculated using the differential latchup criterion. An intermediate excitation time requires a larger bypass current.

Two-dimensional numerical simulations have illustrated the transient behavior of switching from the blocking to the latched state. In one investigation the VPNP base is pulsed, in another the LNPN emitter. As soon as the injected charge is collected, the potential distribution changes dramatically. When the pulse is terminated after the switching current is attained, total current is seen to rise and forward bias the well/substrate junction.

## 7.3 Characterization

The first of two instruments used for most static latchup measurements is the curve tracer. Characterization consists of (1) ungated triggering using supply overvoltage and (2) gated triggering using the curve tracer's base drive to supply a current source to the bypass resistor's terminal. The former measurement is straightforward and has been used extensively. The latter measurement is usually made with the undriven transistor's emitter and bypass resistor shorted together, and the resulting family of characteristics resembles a family of bipolar output characteristics. A sharp upturn (or snapback, depending on load) on the uppermost trace signals latchup. Because of the forward and return trace, the PNPN device is switched in both directions, and the entire I/V characteristic is displayed.

However, a caveat accompanies measurements using a current source in series with the bypass resistance since the tetrode configuration is changed to a triode (or a triode could be changed to a diode). Switching and holding currents measured for the triode configuration are not valid representations for the tetrode if the smaller bypass resistor is in series with the current source. Measurements made with the base drive on opposite bypass resistors produce different results because their trajectories to the switching boundary are in opposite corners of SAFE space and are governed by a different set of device parameters.

The second major instrument is the parameter analyzer, which has more versatile measurement and display capabilities than does the curve tracer. It can simultaneously excite and measure at each of four terminals. Measurements of switching and holding points are more accurate because the I/V characteristic can be displayed, and points can be digitally read directly from the plot. The preferred PNPN characterization technique uses a current source in an emitter leg because emitter current flows in only one direction over the entire range of characteristics. The switching point for the PNPN configuration being measured then corresponds to the experimental point at which $dV_e/dI = 0$.

Various experimental techniques have been used to characterize latchup. Supply overvoltage triggering is commonly used to test for "weak points" in a design, to compare various process options for latchup hardness, or to simply monitor a process. The bypass current causing latchup is junction avalanche, punchthrough, or field FET current. In this technique each emitter is connected to its bypass resistor, and the resulting terminal presented to the power supply, which makes it a two terminal measurement.

There are many experimental techniques for three and four terminal characterization, but some are more useful than others. A current source in series with a bypass resistor alters the PNPN configuration and does not allow current reversal. A voltage source on a bypass resistor or on an emitter causes latchup to occur before it would for a normal tetrode because the applied forward bias augments the bypass current's ohmic drop. Applying a negative bias to a p-type substrate retards latchup, but applying

# Summary

a positive bias enhances latchup. When a current source is used in the emitter leg, the PNPN device exhibits negative feedback at the switching point since the emitter voltage can move up or down to maintain a fixed emitter current. This latter technique displays a portion of the negative differential resistance region when the response time of the measuring instrument is sufficiently short.

Switching point characterization using a current source on the emitter affirms that inverse switching current varies nearly linearly with bypass resistance. (A slightly superlinear behavior is clearly visible and follows from the dependence of critical base/emitter forward bias on bypass resistance.) Measured switching currents are in excellent agreement with the differential latchup criterion for bypass resistance ranging over many orders of magnitude. For equal values of bypass resistance, switching current is lower when the VPNP emitter current is sourced because of its higher current gain and lower emitter/base junction leakage. Lateral current gain often differs from the value measured using the standard 3-terminal technique because holes supplied by the VPNP create an electric field that aids base transport. Although this difference affects the second transistor's emitter current, it has negligible effect on the switching current.

The nearly universal technique for characterizing the holding point is to trigger the PNPN into latchup, then reduce the power supply on the 2-terminal tetrode configuration. Either a voltage source or current source can be used for the power supply, but a portion of the negative differential resistance region is more easily displayed using a current source. Measurements show that inverse holding current also varies nearly linearly with bypass resistance.

Dynamic triggering is used to investigate the transient nature of latchup. Techniques include ramping the power supply voltage, applying a voltage pulse to a pair of substrate or well terminals, and applying a voltage pulse to a bypass resistor terminal. In the first, displacement current flowing through both bypass resistors triggers latchup, and very fast risetimes are required to generate sufficient bypass current. In the second, a pulse of current flows through a single bypass resistor, and a greater pulse height is required as the pulse is shortened. Below some pulse width (typically tens of nanoseconds), latchup does not occur, regardless of pulse height. In the third, an emitter/base junction is forward biased directly by a voltage pulse to simulate I/O over- and undershoot conditions. Again, greater pulse height is required as the pulse is shortened. The critical time for transient latchup is from the start of excitation to the time PNPN current reaches the switching current, although it is not at all clear this is the actual time being reported in the technical literature.

Latchup can be more of a problem on hotter chips. Switching and holding current both decrease at elevated temperature. The forward emitter bias required to generate a particular collector current decreases, and bypass resistance increases with increasing temperature.

Non-electrical probing has also been used for studying latchup. A scanning electron microscope operating in the EBIC mode has been used to identify which circuits on a chip have latched during normal operation or when subjected to a supply overvoltage stress. An infrared microscope has also been used to ascertain sensitive areas by measuring hot spots generated by latchup current. A scanning laser has been used to both trigger and detect latchup on operating circuits.

## 7.4 Avoiding Latchup

There are two fundamental strategies for avoiding latchup.

- Strategy 1. Keep the parasitic PNPN structure in the blocking state.

- Strategy 2. Prevent the parasitic PNPN structure from reaching the latched state.

At first glance these two strategies may appear similar, but there are very important distinctions. The first strategy is implemented by making the switching point inaccessible to any static or dynamic operating point. Usually this means designing the switching current larger than the maximum suppliable current.

The second strategy is implemented by making the holding point inaccessible to any static or dynamic operating point. This means either designing the holding voltage larger than the power supply voltage (plus a safety margin) or designing the holding current larger than the maximum suppliable current. However, in the second case transient SCR behavior can result if device current exceeds the switching current. Behavior of a circuit containing a parasitic PNPN structure can be unpredictable once the PNPN is triggered into the negative differential resistance region. Information can be lost if switching current is exceeded for a sufficiently long time. The safest approach is to avoid entering the negative differential region, even momentarily. This is certainly guaranteed by Strategy 1 and might be accomplished by designing a large enough holding voltage. However, a sufficiently large holding voltage is difficult to design directly for all parasitic PNPN structures encountered in CMOS chip designs because holding voltage is not well modeled. Finally, it should be noted that any of the guidelines that are implemented to raise switching current automatically raise holding current for the same structure, and from equation (4-91) it is clear that for $R_{s2} \neq 0$ raising holding current also raises holding voltage.

Thus, Strategy 1 prevents any PNPN switching, even transient. It is also easier to implement since the switching edge of the blocking state is precisely defined, with no fitting parameters required.

Techniques to implement these strategies divide into two areas - layout guidelines and process design. The first is germane to any CMOS technology and is of interest to the device and circuit designer. The second relates to the process features incorporated into a CMOS technology by the process engineer.

Guard structures are the most effective tool available to the circuit designer. A minority carrier guard is used to pre-collect minority carriers injected into the substrate before they reach a well comprising the base of a vertical bipolar. It consists of a reverse-biased PN junction formed by a source/drain or well diffusion. It is more effective on epi-CMOS because of the reflecting boundary at the high/low junction and because of increased recombination in the highly doped substrate.

A majority carrier guard is used to reduce well or substrate sheet resistance locally, thus minimizing any ohmic drop from parasitic collector current. It consists of a source/drain diffusion of the same type as the background. Operation of a majority guard in the well is also enhanced on epi-CMOS because the elimination of substrate minority carrier collection at the bottom well junction means that majority carriers in the well appear only at its periphery, where they are more easily shunted away from parasitic emitters.

Multiple well contacts augment the low bypass resistance effected by majority guards in the well. Some combination of the two should also be used on I/O circuits to insure the parasitic vertical bipolar is not turned on by I/O over- and undershoots. In addition to avoiding latchup, the goal here is to prevent large current spikes in these high gain devices that otherwise would waste power.

A substrate contact ring should be mandatory for all chips. It minimizes lateral bypass resistance by distributing substrate majority carriers. On epi-CMOS the contact ring can reduce lateral bypass resistance to below 1 $\Omega$.

Butted source contacts can also be used to reduce bypass resistance for the source's parasitic emitter behavior. Their effectiveness in the substrate is greater on epi-CMOS and is improved at reduced epi thicknesses. They are, however, limited to FET's operating with grounded sources.

The second area in which much work has been done to avoid latchup is in process design, and here there are two approaches - bipolar spoiling and bipolar decoupling. Early spoiling techniques attempted to reduce base minority carrier lifetime with either gold doping or neutron irradiation, but this led to adverse device effects such as high leakage current. Later, internal gettering was introduced to lower lifetime in the substrate below the denuded zone, but this has little effect on parasitic lateral bipolars with small basewidth. A retrograde doping profile in the well helps prevent latchup by reducing vertical transport, but care must be taken to remove injected carriers so that lateral transistor action to the well edge is not enhanced. Schottky source/drains on the

# Summary

PMOSFET have been used to degrade emitter injection efficiency, but here a fundamental tradeoff arises between spoiling injection efficiency and maintaining decent FET behavior.

Bipolar decoupling has proved more effective and easier to implement than bipolar spoiling. Several methods are possible. A highly doped substrate beneath a lightly doped epitaxial layer very effectively shunts the lateral parasitic bipolar. As discussed above, this combination also improves the efficacy of minority guards in the substrate and majority guards in the well. A retrograde well can also be used to reduce the well's sheet resistance, although this is usually not as effective as including majority carrier guards in the well. Reverse bias on the substrate (or well) raises the bypass current needed to turn on the corresponding bipolar, but using on-chip generators to provide this bias necessitates careful guard designs to eliminate charge injected by the generator. Trench isolation eliminates lateral current flow to and from the well and conceivably could allow a lithographically limited $N^+/P^+$ spacing. Present sidewall inversion problems must be solved before diffusions can be butted against trench walls, however.

Bipolar decoupling provides a simple design procedure for avoiding latchup. (1) Decouple the vertical parasitic bipolar using epi-CMOS with a substrate contact ring to minimize lateral bypass resistance. (2) Decouple the lateral parasitic bipolar using majority carrier guards in the well or minority carrier guards in the substrate. This design procedure is an important implementation of Strategy 1, an implementation that can be rigorously tested using the differential latchup criterion.

Whichever techniques are chosen to implement Strategy 1, they must prevent switching from the blocking state. Operating strictly in SAFE space guarantees a healthy, latchup-free CMOS technology.

# Appendix A

## Stability Considerations for PNPN Current-Voltage Measurements

In characterizing the switching and holding points with the parameter analyzer, it is often possible to measure a portion of the negative differential resistance region in the vicinity of these points and accurately establish the switching (or holding) current and voltage. This appendix explores the conditions that permit measurements in the negative differential resistance region.

Some authors refer to the lowest point observable in the latched state as the sustaining point (see, for example, [Fang-84]). Other authors have attributed the negative differential resistance region between the holding point and sustaining point to field-aiding effects in the lateral transistor (see, for example, [Rung-82a] and [Rung-83]). However, observations of this region actually depend on the response time of the measuring apparatus and on the resistance in series with the voltage source at the anode. Thus, the current/voltage characteristic between the switching and holding points is not as fundamental to an understanding of latchup as either the characteristic up to and including the switching current or the characteristic for the holding current and above.

Figure A-1 gives the circuit schematic when the PNPN device is excited by a current source. The capacitance C represents the total of current source output capacitance, wiring capacitance, and PNPN terminal capacitance. For convenience, we shall take $I_o$ to be the current at which $dV/dI = 0$, i.e., either the holding current or an excellent approximation to the switching current. This current source is stepped in $\Delta I_o$ increments. The plus sign is taken when measuring the switching current by increasing current through the blocking region and into the negative differential resistance region, and the negative sign is taken when measuring the holding point by decreasing the current through the latched state and into the negative differential resistance region.

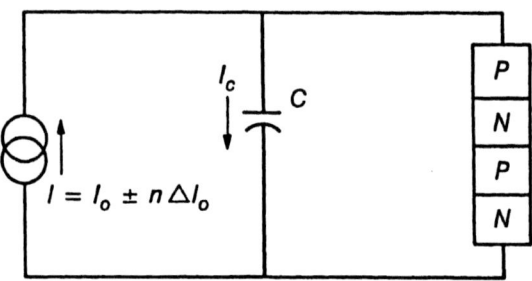

Figure A-1. Circuit schematic for current excitation of PNPN device.

Figure A-2 illustrates the waveform for the current stimulus and corresponding changes on the PNPN current/voltage characteristics; (a) and (b) illustrate measurements near the switching point, (c) and (d), near the holding point. At each step when measuring the switching point, the current source increases by $\Delta I_o$ in a risetime $t_r$. Once current exceeds switching current, terminal voltage decreases by an amount $\Delta V_n = V_n - V_{n-1}$, and the negative differential resistance can be approximated by $r_n = \Delta V_n / \Delta I_o$. Likewise, at each step when measuring the holding point, current decreases by $\Delta I_o$ in a falltime $t_f$. The negative differential resistance is then approximated the same way.

Measuring the switching point causes a current source increase from $I_{n-1} = I_s + (n-1)\Delta I_o$ to $I_n = I_s + n\Delta I_o$ through the effective differential resistance $r_n$. In the absence of capacitor C the resulting equilibrium point would be $(V_n, I_n)$, and theoretically it would then be possible to trace out the entire negative differential resistance region. However, with the capacitance present, the change in terminal voltage causes a current $I_c = C\Delta V_n / t_r$ to flow through the capacitor, in a direction opposite to that shown in Figure A-1 since $\Delta V_n < 0$ for currents just above the switching current. This increases the terminal current flowing into the PNPN device by an amount $-I_c$ and causes an additional reduction in terminal voltage by an amount $I_c r_{n+1}$. If this second voltage reduction is as large as the first, which was produced by the change in the external source, the loop gain for the positive feedback between the capacitor and the negative differential resistance is unity. Thus, an instability results when $I_c r_{n+1} \geq V_n - V_{n-1}$. Substituting from above simplifies this condition to $r_{n+1} C \geq t_r$.

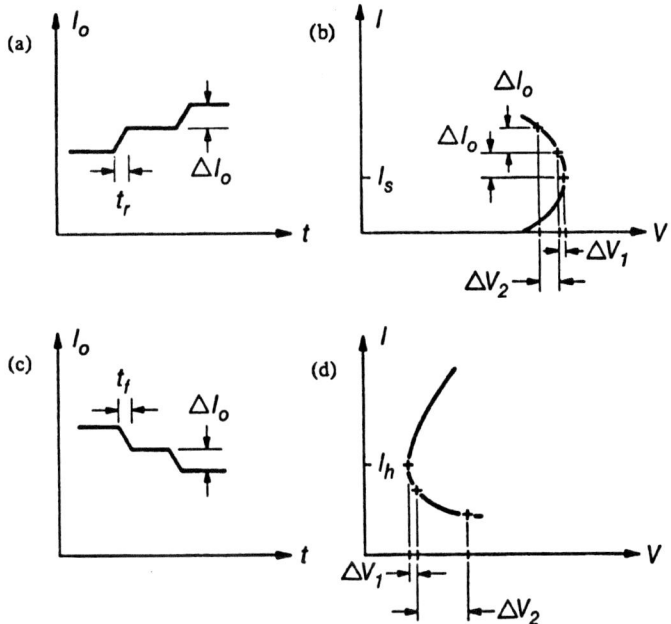

Figure A-2. Illustration of negative differential resistance measurements. (a) Current waveform for measurements near switching point. (b) Negative differential resistance near switching point. (c) Current waveform for measurements near holding point. (d) Negative differential resistance near holding point.

If, on the other hand, the second voltage change is small compared to the first, the additional transient will decay away, and the equilibrium point will be $(V_n, I_n)$. This behavior is consistent with the assumptions made for the current excitation waveform and its transient response. We have tacitly assumed that the time constant $r_n C < \Delta t$ so that the circuit has completed its response to one step change before another step occurs.

The same analysis for measurements at the holding point yield the same result as above except fall time $t_f$ replaces rise time $t_r$.

In summary, we have seen that using a current source to trace out the negative differential resistance region will sucessfully display the curve as long as the magnitude of the differential resistance $r < \tau/C$, where $\tau = t_r$ when current is increased from the switching current, and $\tau = t_f$ when current is decreased from the holding point. More of the negative differential resistance region can be viewed for larger $\tau$ (more gradual changes in the current flow) and for smaller capacitance.

# Appendix B

## Possible Latchup Characterization Problems

The HP-4145A Parameter Analyzer is becoming popular for characterizing latchup because of its versatility and ease of use. There are potential problems, however. Because the PNPN device can be biased to provide positive feedback, the tester itself can sometimes cause seemingly unpredictable latchup behavior. The unwary user can be led to erroneous conclusions about latchup behavior. Below are two common problems and how to avoid them.

### B.1 Problem 1: Triggering from SMU Noise

The first problem arises from the internal workings of the tester's system measurement units (SMU's). The problem occurs when all three of the following occur:

(1) An SMU is holding a voltage at node N.
(2) The same SMU is measuring current at node N.
(3) The PNPN device under test would latch if node N were placed on ground.

When the current meter of the SMU autoranges, the feedback resistor in the SMU's operational amplifier is momentarily open circuited. The input (connected to node N) then tends toward the lower rail voltage (ground). This noise is momentary, but if it persists long enough, it can cause latchup.

If one attempts to measure bypass current $I_{rw}$ while holding both the N-well and $P^+$ emitter terminals to 5 volts, the autoranging feature of the ammeter causes the N-well terminal to momentarily shift toward ground when the current range changes. If the resultant drop in voltage from 5 volts exceeds approximately 0.5 volt, the PNP is turned on. If the resulting PNP collector current is large enough to then turn on the NPN, latchup results. Figure B-1 shows the measured substrate potential at the switching point when current is measured at one, two, three, or all four terminals. SMU's at PNP emitter and N-well are programmed to be 5 volts, and NPN emitter and substrate are programmed to be zero volts.

| Iep | Irw | Ien | Irs | Vs |
|---|---|---|---|---|
| X | | | | 0.80 |
| | X | | | 0.48 |
| | | X | | 0.80 |
| | | | X | 0.80 |
| X | X | | | 0.41 |
| X | | X | | 0.80 |
| X | | | X | 0.80 |
| | X | X | | 0.75 |
| | X | | X | 0.27 |
| | | X | X | 0.80 |
| | X | X | X | 0.75 |
| X | | X | X | 0.80 |
| X | X | | X | 0.34 |
| X | X | X | | 0.75 |
| X | X | X | X | 0.75 |

Figure B-1. Substrate potential at the switching point.

The correct value for the critical voltage of $V_s = 0.80$V is obtained only when $I_{rw}$ is not measured. The other values are lower because the VPNP is being triggered in those cases as well as the LNPN. The easiest way to avoid the problem is not to measure the current at the node to the bypass resistor if that node is being held at any potential other than ground. For the 4-terminal PNPN structure $I_{rw}$ can be calculated by summing the measured values of the other three currents.

## B.2 Problem 2: SMU Output Sequence

During testing terminal voltages and currents are asserted one node at a time. The default order follows the numbering of the SMU'S, i.e., SMU 1 is asserted first, then SMU 2, etc., but the output sequence set-up screen on the HP-4145A allows the user to redefine the order in which the SMU's are asserted. For many measurements this order has no effect on the results, but in doing latchup measurements on potentially high gain devices the user must be careful not to latch the device by using an improper assertion order. Since an SMU behaves like an open circuit until a particular voltage or current is asserted, asserting the SMU's on the anode and cathode terminals before the SMU's on the well and substrate terminals momentarily results in a Shockley diode configuration. If the sum of transistor alphas exceeds unity, the result is immediate latchup. Once the device is in the low impedance state, it will remain there even after the remaining two SMU's are asserted.

# References

References are listed chronologically with the year of publication included in the citation.

[Shockley-50] W. Shockley, "Theories of High Values of Alpha for Collector Contacts on Germanium," Phys. Rev., vol. 78, pp. 294-295 (April 15, 1950).

[Shockley-51] W. Shockley, M. Sparks, and G. K. Teal, "p-n Junctions Transistors," Phys. Rev., vol. 83, pp. 151-162 (July 1, 1951).

[Ebers-52] J.J. Ebers, "Four-terminal pnpn transistors," Proc. IRE, vol. 40, pp. 1361-1364 (Nov. 1952).

[Moll-56] J.L. Moll, M. Tannenbaum, J.M. Goldey, and N. Holonyak, "pnpn transistors switches," Proc. IRE, vol. 44, pp. 1174-1182 (Sept. 1956).

[Mackintosh-58] I.M. Mackintosh, "The Electrical Characteristics of Silicon PNPN Triodes," Proc. IRE, vol. 46, pp. 1229-1235 (June 1958).

[Wanlass-63] F.M. Wanlass and C.T. Sah, "Nanowatt Logic Using Field-Effect Metal Oxide Semiconductor Triodes," 1963 Int. Solid State Circuits Conf., pp. 32-33 (Feb. 1963).

[Gentry-64] F.E. Gentry, F.W. Gutzwiller, N. Holonyak, and E.E. Von Zastrow, *Semiconductor Controlled Rectifiers*, Englewood Cliffs, N.J., Prentice-Hall (1964).

[Gibbons-64] J.F. Gibbons, "A critique of the theory of pnpn devices," IEEE Trans. Elec. Dev., vol.ED-11, pp. 406-413 (Sept. 1964).

[Lindmayer-65] J. Lindmayer and C.Y. Wrigley, *Fundamentals of Semiconductor Devices*, D. Van Nostrand Co., Inc. Princeton, N. J. (1965).

[Gregory-73] B.L. Gregory and B.D. Shafer, "Latch-up in CMOS integrated circuits," IEEE Trans. Nucl. Sci., vol. NS-20, pp. 293-299 (Dec. 1973).

[Kennedy-73] D. P. Kennedy and A. Phillips, Jr., "Source-Drain Breakdown in an Insulated Gate Field-Effect Transistor," 1973 IEDM Technical Digest, pp. 160-163 (Dec. 1973).

[Dawes-76] W. R. Dawes, Jr. and G. F. Derbenwick, "Prevention of CMOS Latchup in MOS Integrated Circuits," IEEE Trans. Nucl. Sci., NS-23, pp. 2027-2030 (Dec. 1976).

[Getreau-78] I. E. Getreau, *Modeling the Bipolar Transistor*, Elsevier Scientific, New York (1978).

[Kyomasu-78] M. Kyomasu, T. Araki, T. Ohtsuki, and M. Nakayama, "Analysis of latchup phemomena in CMOS IC," Trans. IECE Japan, vol. E61, no. 2, pp. 109-110 (Feb. 1978).

[Estreich-78] D.B. Estreich, A. Ochoa, Jr., and R.W. Dutton, "An Analysis of Latch-up Prevention in CMOS IC's Using an Epitaxial-Buried Layer Process," 1978 IEDM Technical Digest, pp. 230-234 (Dec. 1978).

[Ochoa-79] A. Ochoa, W. Dawes, and D. Estreich, "Latch-up control in integrated circuits," IEEE Trans. Nucl. Sci., vol. NS-26, pp. 5065-5068 (Dec. 1979).

[Adams-79] J. R. Adams and J. R. Sokel, "Neutron Irradiation for Prevention of Latchup in MOS Integrated Circuits," IEEE Trans. Nucl. Sci., NS-26, pp. 5069-5073 (Dec. 1979).

# References

[Estreich-80] D. B. Estreich, "The Physics and Modeling of Latch-up and CMOS Integrated Circuits," Tech. Report No. G-201-9, Stanford Electronics Lab., Stanford Univ., (Nov. 1980).

[Dressendorfer-80] P. V. Dressendorfer and M. G. Armendariz, "A SEM Technique for Experimentally Locating Latchup Paths in Integrated Circuits," IEEE Trans. Nucl. Sci., NS-27, pp. 1688-1693 (Dec. 1980).

[Schroeder-80] J.E. Schroeder, A. Ochoa, and P.V. Dressendorfer," Latch-up Elimination in Bulk CMOS LSI Circuits," IEEE Trans. Nucl. Sci., vol. NS-27, pp.1735-1738 (Dec. 1980).

[Payne-80] R. S Payne, W. R. Grant, and W. J. Bertram, "Elimination of Latchup in Bulk CMOS," 1980 IEDM Technical Digest, pp. 248-251 (Dec. 1980).

[Raburn-80] W.D. Raburn, "A model for the parasitic SCR in bulk CMOS," IEDM Tech. Dig., pp. 252-255 (Dec. 1980).

[Parrillo-80] L. C. Parrillo, R. S. Payne, R. E. Davis, G. W. Reutlinger, and R. L. Field, "Twin-Tub CMOS - A Technology for VLSI Circuits," 1980 IEDM Technical Digest, pp. 752-755 (Dec. 1980).

[Rung-81] R. D. Rung, C. J. Dell'Oca, and L. G. Walker, "A Retrograde P-well for Higher Density CMOS," IEEE Trans. Elec. Dev., ED-28, pp. 1115-1119 (Dec. 1981).

[Iizuka-81] T. Iizuka and J. L. Moll, "A Figure of Merit for CMOS Latchup Tolerance," 1981 CMOS Workshop, San Francisco, Ca. (May 18, 1981).

[Dressendorfer-81] P.V. Dressendorfer and A. Ochoa, Jr., "An Analysis of the Modes of Operation of Parasitic IC's," IEEE Trans. Nucl. Sci., vol. NS-28, pp.4288-4291 (Dec. 1981).

[Ochoa-81] A. Ochoa and P.V. Dressendorfer, "A discussion of the role of distributed effects in latch-up," IEEE Trans. Nucl. Sci., vol. NS-28, pp. 4292-4294 (Dec. 1981).

[Combs-81] S. R. Combs, "Scaleable Retrograde P-well CMOS Technology," 1981 IEDM Technical Digest, pp.346-349 (Dec. 1981).

[Wieder-81] A. W. Wieder, C. Werner, and J. Harter, "Design Model for Bulk CMOS Scaling Enabling Accurate Latchup Prediction," 1981 IEDM Technical Digest, pp. 354-358 (Dec. 1981).

[Sakai-81] Y. Sakai, T. Hayashida, N. Hashimoto, O. Mimato, T. Masuhara, K. Nagasawa, T.Yasui, and N. Tanimara, "Advanced Hi-CMOS Device Technology," 1981 IEDM Technical Digest, pp. 534-537 (Dec. 1981).

[Hu-82a] G. J. Hu, M. R. Pinto, and S. Kordic, "Two-Dimensional Simulation of Latch-Up in CMOS Structure," paper VA-5, 40th Device Research Conf., Colorado State Univ., Ft. Collins, Colo. (June 21-23, 1982).

[Rung-82a] R. D. Rung and H. Momose, "Improved Modeling of CMOS Latch-Up and VLSI Implications," Digest of 1982 Symposium on VLSI Technology, pp.50-51 (Sept. 1982).

[Troutman-82] R. R. Troutman and H. P. Zappe, "Power-Up Triggering Conditions for Latchup in Bulk CMOS," Digest of 1982 Symposium on VLSI Technology, pp. 52-53 (Sept. 1982).

[Rung-82b] R. D. Rung, H. Momose, and Y. Nagakubo, "Deep Trench Isolated CMOS Devices," 1982 IEDM Technical Digest, pp. 237-240 (Dec. 1982).

[Jerdonek-82] R. Jerdonek, M. Ghezzo, J. Weaver, and S. Combs, "Reduced Geometry CMOS Technology," 1982 IEDM Technical Digest, pp. 450-453 (Dec. 1982).

[Huang-82] C. C. Huang, M. D. Hartranft. N. F. Pu, C. Yue, C. Rahn, J. Schrankler, G. D. Kirchner, F. L. Hampton, and T. E. Hendrickson, "Characterization of CMOS Latchup," 1982 IEDM Technical Digest, pp. 454-457 (Dec. 1982).

# References

[Takacs-82] D. Takacs, C. Werner, J. Harter, and U. Schwabe, "Surface Induced Latch-Up in VLSI CMOS Circuits," 1982 IEDM Technical Digest, pp. 458-461 (Dec. 1982).

[Sugino-82] M. Sugino, L.A. Akers, and M. E. Rebeschini, "CMOS Latch-up Elimination Using Schottky Barrier PMOS," 1982 IEDM Technical Digest, pp. 462-465 (Dec. 1982).

[Koeneke-82] C. J. Koeneke and W. T. Lynch, "Lightly Doped Schottky MOSFET," 1982 IEDM Technical Digest, pp. 466-469 (Dec. 1982).

[Hashimoto-82] K. Hashimoto, S. Morita, H. Nozawa, and S. Kohyama, "Counterdoped Well Structure for Scaled CMOS," 1982 IEDM Technical Digest, pp. 470-473 (Dec. 1982).

[Hu-82b] G. J. Hu, C. Y. Ting, Y. Taur, and R. H. Dennard, "Design and Fabrication of P-channel FET's for 1 $\mu$m CMOS Technology," 1982 IEDM Technical Digest, pp. 710-713 (Dec. 1982).

[Sugino-83] M. Sugino, L. A. Akers, and M. E. Rebeschini, "Latchup-Free Schottky-Barrier CMOS," IEEE Trans. Elec. Dev., ED-30, pp. 110-118 (Feb. 1983).

[Troutman-83a] R. R. Troutman and H. P. Zappe, "A Transient Analysis of Latchup in Bulk CMOS," IEEE Trans. Elec. Dev., ED-30, pp. 170-179 (Feb. 1983).

[Wieder-83] A. W. Wieder, C. Werner, and J. Harter, "Design Model for Bulk CMOS Scaling Enabling Accurate Prediction," 1981 IEDM Digest, pp. 354-358 (Dec. 1981). Also see IEEE Trans. on Elec. Devices, ED-30, pp.240-245 (March 1983).

[Manoliu-83] J. Manoliu, F. H. Tseng, B. J. Woo, and T. J. Meier, "High Density and Reduced Latchup Susceptibility CMOS Technology for VLSI," IEEE Electron Dev. Letters, EDL-4, pp. 233-235 (July 1983).

[Wakeman-83] L. Wakeman, "Silicon-gate CMOS chips gain immunity to SCR latchup," Electronics, pp. 136-140 (Aug. 11, 1983).

[Troutman-83b] R. R. Troutman and M. J. Hargrove, "Transmission Line Model for Latchup in CMOS Circuits," Digest of Technical Papers for 1983 Symposium on VLSI Technology, pp. 56-59 (Sept. 13-15, 1983).

[Cham-83] K. M. Cham and S.-Y. Chiang, "A Study of the Trench Surface Inversion Problem in the Trench CMOS Technology," IEEE Elec. Dev. Letters, EDL-4, pp. 303-305 (Sept. 1983).

[Davies-83] R. D. Davies, "The Case for CMOS," IEEE Spectrum, pp. 26-32 (Oct. 1983).

[Schwabe-83] U. Schwabe, H. Herbst, E. J. Jacobs, and D. Takacs, "N- and P-well Optimization for High-Speed N-Epitaxy CMOS Circuits," IEEE Trans. Elec. Dev., ED-30, pp. 1339-1344 (Oct. 1983).

[Troutman-83c] R. R. Troutman, "Epitaxial Layer Enhancement of n-Well Guard Rings for CMOS Circuits," IEEE Elec. Dev. Letters, vol. ED-4, pp. 438-440 (Dec. 1983).

[Rung-83] R. D. Rung and H. Momose, "DC Holding and Dynamic Triggering Characteristics of Bulk CMOS Latchup," IEEE Trans. Elec. Dev., vol. ED-30. pp. 1647-1655 (Dec. 1983).

[Endo-83] N. Endo, N Kasai, A. Ishitani, and Y. Kurogi, "CMOS Technology Using SEG Isolation Techniques," 1983 IEDM Technical Digest, pp. 31-34 (Dec. 1983).

[Wollesen-83] D. L. Wollesen, J. Haskell, and J. Yu, "N-well and P-well Performance comparison," 1983 IEDM Technical Digest, pp. 155-158 (Dec. 1983).

[Takacs-83] D. Takacs, J. Harter, E. P. Jacobs, C. Werner, U. Schwabe, J. Winnerl, and E. Lange, "Comparison of latch-up in

# References

N-well and P-well CMOS circuits," 1983 IEDM Technical Digest, pp. 159-164 (Dec. 1983).

[Niitsu-83] Y. Niitsu, H. Nihira, K. Kanzaki, and S. Kohyama, "Resistance Modulation Effect Due to current injection and CMOS latch-up," 1983 IEDM Technical Digest, pp. 164-167 (Dec. 1983).

[Goto-83] G. Goto, H. Takahashi, and T. Nakamura, "Latch-up Immunity against noise pulses in a CMOS double well structure," 1983 IEDM Technical Digest, pp. 168-171 (Dec 1983).

[Hamdy-83] E. Hamdy and A. Mohsen, "Characterization and modeling of transient latchup in CHMOS Technology," 1983 IEDM Technical Digest, pp. 172-175 (Dec. 1983).

[Hu-83] G. J. Hu, Y. Taur, R. H. Dennard, L. M. Terman, and C. Y. Ting, "A Self-Aligned 1 $\mu$m CMOS Technology for VLSI," 1983 IEDM Technical Digest, p. 731 (Dec. 1983).

[Hu-84a] G. J. Hu, "A Better Understanding of CMOS Latch-Up," IEEE Trans. Elec. Dev., vol. ED-31, pp. 62-67 (Jan. 1984).

[Fang-84] R. C. Fang and J. L. Moll, "Latchup Model for the Parasitic Path in Bulk CMOS," IEEE Trans. Elec. Dev.,vol. ED-31, pp. 113-120 (Jan. 1984).

[Dooley-84] J. G. Dooley and R. C. Jaeger, "Temperature Dependence of Latchup in CMOS Circuits," IEEE Elec. Dev. Letters, EDL-5, pp. 41-43 (Feb. 1984).

[Anagnostopoulos-84] C. N. Anagnostopoulos, E. T. Nelson, J. P. Levine, K. Y. Wong, and N. Nichols, "Latchup and Image Crosstalk Suppression by Internal Gettering," IEEE J. Solid-State Circuits, SC-19, pp. 91-97 (Feb. 1984).

[Craig-84] W. Craig, "Latchup Test Structures and Their Characterization," IEEE VLSI Workshop on Test Structures, Session IV, San Diego, Ca. (Feb. 20-21, 1984).

[Takacs-84] D. Takacs, C. Werner, J. Harter, and U. Schwabe, "Surface Induced Latchup in VLSI CMOS Circuits," IEEE Trans. Elec. Dev., vol. ED-31, pp. 279-286 (March 1984).

[Troutman-84] R. R. Troutman and H. P Zappe, "Layout and Bias Considerations for Preventing Transiently Triggered Latchup in CMOS," IEEE Trans. Elec. Dev., ED-31, pp. 315-321 (March 1984).

[Kokkonen-84] K. Kokkonen and R. Pashley, "Modular approach to CMOS tailors process to application," Electronics, pp. 129-133 (May 3, 1984).

[Leventhal-84] I. H. Leventhal, "Comparison of DC Latchup Characterization Techniques for CMOS Technology," BS Thesis, MIT (June 1984).

[Hu-84b] G. J. Hu and R. H. Bruce, "A CMOS Structure with High Latchup Holding Voltage," IEEE Elec. Dev. Letters, EDL-5, pp. 211-214 (June 1984).

[Sangiori-84] E. Sangiori and S. Swirhun, "Trenched Schottky Barrier PMOS for Latchup Resistance," IEEE Elec. Dev. Letters, pp. 293-295 (Aug. 1984).

[Terrill-84a] K. W. Terrill and C. Hu, "Substrate Resistance Calculation for Latchup Modeling," IEEE Trans. Elec. Dev., ED-31, pp. 1152-1155 (Sept. 1984).

[Lewis-84] A. G. Lewis, "Latchup Suppression in Fine-Dimension Shallow P-well CMOS Circuits," IEEE Trans. Elec. Dev., pp. 1472-1481 (Oct. 1984).

[Taur-84] Y. Taur, W. H. Chang, and R. H. Dennard, "Characterization and Modeling of a Latchup-Free 1-$\mu$m CMOS Technology," 1984 IEDM Technical Digest, pp. 398-401 (Dec. 1984).

[Swirhun-84] S. Swirhun, E. Sangiorgi, A. Weeks, R. M. Swanson, K. C. Saraswat, and R. W. Dutton, "Latchup Free

# References

CMOS Using Guarded Schottky Barrier PMOS," 1984 IEDM Technical Digest, pp. 402-405 (Dec. 1984).

[Terrill-84b] K. W. Terrill, P. F. Byrne, H. P. Zappe, N. W. Cheung, and C. Hu, "A New Method for Preventing CMOS Latchup," 1984 IEDM Technical Digest, pp. 406-409 (Dec. 1984).

[Schwabe-84] U. Schwabe, E. P. Jacobs, D. Takacs, J. Winnerl, and E. Lange, "Reduced $N^+/P^+$ Spacing with High Latchup Hardness in Self-Aligned Double Well CMOS Technology," 1984 IEDM Technical Digest, pp. 410-413 (Dec. 1984).

[Pinto-85] M. R. Pinto and R. W. Dutton, "Accurate Triggering Condition Analysis for CMOS Latchup," IEEE Electron Dev. Letters, pp. 100-102 (Feb. 1985).

[Fu-85] K. Y. Fu, "Transient Latchup in Bulk CMOS with a Voltage-Dependent Well-Substrate Junction Capacitance," IEEE Trans. Elec. Dev., ED-32, pp. 717-720 (March 1985).

[Hall-85] J. E. Hall, J. A. Seitchik, L. A. Arledge, and P. Yang, "An Improved Circuit Model for CMOS Latchup," IEEE Elec. Dev. Letters, EDL-6, pp. 320-322 (July 1985).

[Odanaka-85] S. Odanaka, M. Wakabayashi, and T. Ohzone, "The Dynamics of Latchup Turn-On Behavior in Scaled CMOS," IEEE Trans. Elec. Dev., ED-32, pp. 1334-1340 (July 1985).

[Piro-85] R. Piro and F. Sporck, "Latchup-Free Substrate Bias Generators in CMOS," IEEE Custom Integrated Circuits Conf. Digest, pp. 524-527 (May 1985).

[Troutman-86] R. R. Troutman and M. J. Hargrove, "Transmission Line Modeling of Substrate Resistance and CMOS Latchup," IEEE Trans. Elec. Dev. (July 1986).

# Glossary

## SYMBOL DEFINITIONS

$ALU$     Arithemetic logic unit.

$\alpha_{fn}$     Large-signal, forward NPN common-base current gain.

$\alpha_{fn}^*$     Effective large-signal, forward NPN common-base current gain; includes effect of NPN bypass resistance.

$\alpha_{fns}$     Small-signal, forward NPN common-base current gain.

$\alpha_{fns}^*$     Effective small-signal, forward NPN common-base current gain; includes effect of NPN bypass resistance.

$\alpha_{fp}$     Large-signal, forward PNP common-base current gain.

$\alpha_{fp}^*$     Effective large-signal, forward PNP common-base current gain; includes effect of PNP bypass resistance.

$\alpha_{fps}$     Small-signal, forward PNP common-base current gain.

$\alpha_{fps}^*$     Effective small-signal, forward PNP common-base current gain; includes effect of PNP bypass resistance.

$\alpha_{rn}$     Large-signal, reverse NPN common-base current gain.

$\alpha_{rns}$     Small-signal, reverse NPN common-base current gain.

$\alpha_{rp}$     Large-signal, reverse PNP common-base current gain.

$\alpha_{rps}$     Small-signal, reverse PNP common-base current gain.

| | |
|---|---|
| $\beta_{fn}$ | Large-signal, forward NPN common-emitter current gain. |
| $\beta_{fns}^*$ | Small-signal, forward NPN common-emitter current gain. |
| $\beta_{fp}$ | Large-signal, forward PNP common-emitter current gain. |
| $\beta_{fps}$ | Small-signal, forward PNP common-emitter current gain. |
| $DLC$ | Differential latchup criterion. |
| $DLP$ | Decouple lateral parasitic. |
| $D_n$ | Diffusion constant for electrons in neutral NPN base. |
| $D_p$ | Diffusion constant for electrons in neutral PNP base. |
| $DVP$ | Decouple vertical parasitic. |
| $EBIC$ | Electron beam induced current. |
| $G_{bn}$ | Small-signal NPN base junction conductance. |
| $\gamma$ | Characteristic length of transmission line. |
| $\gamma_n^*$ | Large-signal, effective NPN injection efficiency. See eq. 4-20. |
| $\gamma_{ns}^*$ | Small-signal, effective NPN injection efficiency. See eq. 4-20. |
| $\gamma_p^*$ | Large-signal, effective PNP injection efficiency. See eq. 4-19. |

# Symbol Definitions

$\gamma_{ps}^*$      Small-signal, effective PNP injection efficiency. See eq. 4-19.

$H_n$      NPN high level injection factor. See eq. 4-45.

$H_p$      PNP high level injection factor.

$I_a$      Anode current.

$I_{avb}$      Bulk avalanche current. See Fig. 3-1.

$I_{avs}$      Surface avalanche current. See Fig. 3-1.

$I_{bn}$      NPN base current.

$I_{bp}$      PNP base current.

$I_{by}$      Parasitic bipolar bypass current.

$I_c$      Bipolar collector current. PNPN cathode current.

$I_{ci}$      Parasitic vertical transistor's collector current collected at side of well junction.

$I_{cn}$      NPN collector current.

$I_{co}$      Parasitic vertical transistor's collector current measured at topside substrate contact.

$I_{cp}$      PNP collector current.

$I_{crit}$      Critical current (sometimes used to label switching current).

$I_{Dmax}$      Maximum power supply current.

$I_e$      Bipolar emitter current.

$I_{en}$      NPN emitter current.

| | |
|---|---|
| $I_{en,h}$ | NPN emitter current at the holding point. |
| $I_{eno}$ | NPN emitter current at low level injection. |
| $I_{en,s}$ | NPN emitter current at the switching current when parasitic PNPN is excited by PNP. |
| $I_{en,so}$ | Current $I_{en,s}$ in the absence of high level injection effects. |
| $I_{ep}$ | PNP emitter current. |
| $I_{ep,h}$ | PNP emitter current at the holding point. |
| $I_{ep,s}$ | PNP emitter current at the switching current when parasitic PNPN is excited by NPN. |
| $I_{ep,so}$ | Current $I_{ep,s}$ in the absence of high level injection effects. |
| $I_g$ | PNPN gate current. |
| $I_h$ | Holding current. |
| $I_j$ | Reverse biased PNPN center junction current. |
| $I_{kfn}$ | Knee current (on plot of log $I_{cn}$ vs. $V_{ben}$) for NPN operating in forward mode. |
| $I_{kfp}$ | Knee current (on plot of log $I_{cp}$ vs. $V_{ebp}$) for PNP operating in forward mode. |
| $I_n$ | Total electron current at PNPN center junction. |
| $I_{nd}$ | Component of electron current due to diffusion. |
| $I_{nsc}$ | Component of electron current due to space charge generation. |

## Symbol Definitions

| | |
|---|---|
| $I_{nw}$ | Current measured at N-well terminal ( $= I_{rw}$). |
| $I_o$ | Well/substrate current source. |
| $I_p$ | Total hole current at PNPN center junction. |
| $I_{pd}$ | Component of hole current due to diffusion. |
| $I_{psc}$ | Component of hole current due to space charge generation. |
| $I_{pts}$ | External punchthrough current. See Fig. 3-3. |
| $I_{pt1}$ | Lateral, internal punchthrough current. See Fig. 3-3. |
| $I_{pt2}$ | Vertical, internal punchthrough current. See Fig. 3-3. |
| $I_r$ | PNPN center junction current. |
| $I_{r,h}$ | PNPN center junction emitter current at the holding point. |
| $I_{rs}$ | Bypass current through resistor $R_s$. |
| $I_{rs,to}$ | Current $I_{rs}$ at the turn-off point. |
| $I_{rw}$ | Bypass current through resistor $R_w$. |
| $I_{rw,to}$ | Current $I_{rw}$ at the turn-off point. |
| $I_s$ | Switching current. |
| $I_{sf}$ | Emitter/base junction saturation current. |
| $I_{sn}$ | NPN base/emitter junction saturation current. |
| $I_{sn,x}$ | Switching current measured when NPN emitter current is sourced. |
| $I_{sp}$ | PNP base/emitter junction saturation current. |

$I_{sp,x}$     Switching current measured when PNP emitter current is sourced.

$I_{sr}$     PNPN collector/base (center) junction saturation current.

$I_{s1}$     Internal diffusion to well junction saturation current.

$I_{s2}$     PNPN center junction saturation current.

$I_{s3}$     External diffusion to substrate junction saturation current.

$I_{to}$     PNPN turn-off current.

$I_{wj}$     Total current at well/substrate junction.

$I_z$     Maximum latched state current. See Fig. 2-4.

$I_{1,on}$     PNPN device current at which first transistor turns on in piecewise linear model.

$I_{2,on}$     PNPN device current at which second transistor turns on in piecewise linear model.

$K$     Ratio of bypass resistance to small-signal emitter resistance.

$\kappa$     Ratio of small-signal base conductance to sum of bypass conductance and small-signal base conductance.

$LNPN$     Lateral NPN transistor.

$L1$     Length of transmission line from N-well to external $N^+$ diffusion. See Fig. 4-15.

$L2$     Length of transmission line from $N^+$ diffusion to topside substrate contact when latter is closer to N-well. See Fig. 4-15.

# Symbol Definitions

$L3$      Length of transmission line from $N^+$ diffusion to topside substrate contact when former is closer to N-well. See Fig. 4-22.

$MLU$      Main latchup.

$M_n$      Electron multiplication factor.

$M_p$      Hole multiplication factor.

$OLU$      Output latchup.

$SEM$      Scanning electron microscope.

$\psi_s$      Surface potential.

$\psi_{sni}$      Surface potential induced by lateral well current in VPNP.

$\psi_{sno}$      Surface potential induced by vertical well current in VPNP.

$q_{bn}$      NPN normalized base charge.

$q_{bp}$      PNP normalized base charge.

$q_{1n}$      NPN basewidth modulation factor.

$q_{2n}$      NPN high level injection factor.

$R_{bn}$      NPN base leg resistance.

$R_{bp}$      PNP base leg resistance.

| | |
|---|---|
| $R_{cn}$ | NPN collector resistance. |
| $R_{cp}$ | PNP collector resistance. |
| $r_{en}$ | Small-signal NPN emitter junction resistance. |
| $R_{en}$ | NPN emitter leg resistance. |
| $R_{en,h}$ | Small-signal resistance $r_{en}$ at the holding point. |
| $r_{ep}$ | Small-signal PNP emitter junction resistance. |
| $R_{ep}$ | PNP emitter leg resistance. |
| $R_{ep,h}$ | Small-signal resistance $r_{ep}$ at the holding point. |
| $R_n$ | Transmission line insertion resistance from $N^+$ diffusion. |
| $R_{nq}$ | Parallel resistance of $r_{en}$ and $R_s$. |
| $R_{nq,h}$ | Resistance $R_{nq}$ at the holding point. |
| $R_p$ | Direct resistance from topside diffusion to highly doped substrate through the epitaxial layer. |
| $R_{pq}$ | Parallel resistance of $r_{ep}$ and $R_w$. |
| $R_{pq,h}$ | Resistance $R_{pq}$ at the holding point. |
| $r_r$ | Small-signal resistance of PNPN center junction. |
| $r_{r,h}$ | Small-signal resistance $r_r$ at the holding point. |
| $R_s$ | Substrate resistance. Lateral bypass resistance for N-well CMOS. |
| $R_{s,eq}$ | Equivalent substrate bypass resistance from transmission line model. |

# Symbol Definitions

| | |
|---|---|
| $R_{si}$ | Internal component of lateral NPN bypass resistance. |
| $R_{sx}$ | External component of lateral NPN bypass resistance. |
| $R_{s1}$ | Lateral NPN bypass resistance. See Fig. 2-6. |
| $R_{s2}$ | Vertical PNP bypass resistance. See Fig. 2-6. |
| $R_{s3}$ | See Fig. 2-2. |
| $R_{s4}$ | See Fig. 2-2. |
| $R_{ti}$ | Input-excited transfer resistance. |
| $R_{to}$ | Output-excited transfer resistance. |
| $R_w$ | Well resistance. |
| $R_{wi}$ | Internal component of vertical PNP bypass resistance. |
| $R_{wx}$ | External component of vertical PNP bypass resistance. |
| $R_{w1}$ | Vertical PNP bypass resistance. |
| $R_{w2}$ | Lateral NPN collector resistance. |
| $R_{w3}$ | See Fig. 2-2. |
| $R_{w4}$ | See Fig. 2-2. |
| $R_x$ | Series resistance external to PNPN device. |
| $R_1$ | (1) Transmission line input resistance. (2) Bypass resistance for first transistor to turn on. |
| $R_2$ | (1) Transmission line output resistance. (2) Bypass resistance for second transistor to turn on. |

| | |
|---|---|
| *SCR* | Semiconductor controlled rectifier. |
| $\tau_n$ | NPN base transit time. |
| $\tau_p$ | PNP base transit time. |
| $\tau_1$ | Time constant characterizing response to ramped power supply (0 transistors turned on). |
| $\tau_2$ | Time constant characterizing response to ramped power supply (1 transistor turned on). |
| $\tau_3$ | Time constant characterizing response to ramped power supply (2 transistors turned on). |
| $V_{bcn}$ | NPN base/collector voltage. |
| $V_{ben}$ | NPN base/emitter voltage. |
| $V_{ben,s}$ | NPN base/emitter voltage at the switching point. |
| $V_{ebp,s}$ | PNP base/emitter voltage at the switching point. |
| $V_{b,on}$ | Emitter/base forward bias at which transistor is turned on in piecewise linear model. |
| $V_{cbp}$ | PNP collector/base voltage. |
| $V_{ce}$ | Collector/emitter voltage. |
| $V_{crit}$ | Critical voltage. |
| $V_{dd}$ | Positive power supply voltage. |
| $V_{ebp}$ | PNP emitter/base voltage. |

## Symbol Definitions

| | |
|---|---|
| $V_h$ | Holding voltage. |
| $V_{in}$ | Inverter input voltage. |
| $V_{out}$ | Inverter output voltage. |
| $VPNP$ | Vertical PNP transistor. |
| $V_r$ | Forward bias on PNPN center junction. |
| $V_s$ | Switching voltage. |
| $V_{ss}$ | Most negative power supply voltage. |
| $V_{sub}$ | Substrate bias. |
| $V_t$ | Thermal voltage. |
| $V_{to}$ | PNPN turn-off voltage. |
| $V_1$ | Parasitic PNP emitter/base forward bias. |
| $V_2$ | PNPN center junction reverse bias. |
| $V_3$ | Parasitic NPN base/emitter forward bias. |
| $W_n$ | NPN base width. |
| $W_p$ | PNP base width. |
| $Z_{in}$ | Transmission line input impedance. |
| $Z_o$ | Transmission line characteristic impedance. |

# Index

Backside substrate contact ... 10, 94
Bipolar decoupling ... 4, 174, 180-190, 192, 207
Bipolar spoiling ... 4, 174-180, 192, 206
Blocking state ... 11, 19, 21, 23, 24, 32, 34, 35, 37, 42, 45, 47, 51, 53, 56, 57, 58, 59, 63, 66, 67, 69, 71, 73, 74, 75, 76, 80, 81, 86, 87, 90, 94, 108, 109, 110, 120, 124, 125, 128, 130, 132, 135, 136, 138, 164, 197, 198, 199, 204, 205, 207
Butted contact ... 10, 152, 173-174
Bypass current ... 13, 18, 34, 35, 45, 52, 53, 54, 56, 63, 68, 74, 77, 78, 83, 85, 86, 89, 132, 134, 135, 136, 147, 153, 163, 197, 198, 201, 202, 203, 207, 213
Bypass resistor ... 13, 24, 26, 32, 33, 34, 35, 40, 47, 48, 51, 52, 54, 57, 58, 63, 67, 68, 69, 70, 71, 72, 73, 76, 79, 80, 82, 83, 85, 86, 92, 93, 99, 107, 117, 123, 125, 126, 129, 131, 132, 133, 135, 138, 139, 140, 143, 144, 145, 148, 149, 155, 157, 160, 162, 163, 173, 191, 193, 194, 195, 197, 199, 201, 202, 203
Characterization of latchup
    current source excitation ... 62, 118, 125, 129, 131, 133, 134, 137, 138, 139, 140, 142, 149, 151, 153, 163, 201, 202, 203, 209, 210, 212
    holding point ... 117, 136, 137, 149-154, 160, 163, 202, 209, 210, 211, 212
    non-electrical probing ... 161-162, 204
    switching point ... 117, 125, 133, 134, 136, 137, 138-149, 163, 202, 203, 210, 211, 214
    voltage source excitation ... 26, 118, 131-133, 150, 152, 202, 209
Curve tracer ... 118-124, 126, 136, 180, 182, 201
Differential latchup criterion ... 17, 59, 60, 62, 63, 67, 68, 69, 72, 74, 86, 91, 99, 115, 116, 117, 132, 133, 138, 145, 149, 199, 200, 201, 203
Dynamic latchup effects ... 51, 104-115

Dynamic recovery ... 107, 108, 109
Dynamic triggering ... 155-160, 203
Effective injection efficiency ... 45, 57, 58, 70
Epi-CMOS ... 2, 4, 9, 63, 93, 123, 156, 166, 167, 168, 170, 171, 173, 176, 177, 181-184, 190, 191, 193, 199, 206, 207
Floating N-well ... 15, 16, 18, 20, 66, 69, 70, 76, 91, 92, 132, 139, 194
Floating substrate ... 66, 71, 72, 75, 76, 92, 122, 144
Gate current ... 42, 43, 44, 45, 46, 53, 57, 120, 198
Gated triggering ... 120, 201
General latchup criteria ... 24, 51
Guard structures ... 4, 103, 104, 156, 160, 165-172, 173, 176, 177, 181, 182, 183, 184, 186, 188, 191, 192, 195, 205, 206, 207
High level injection ... 17, 49, 63, 105, 115, 140, 161, 182, 199
Holding current ... 12, 19, 21, 24, 43, 45, 46, 51, 53, 54, 77, 78, 81-83, 88, 90, 104, 117, 122, 123, 124, 137, 149, 151, 152, 153, 160, 166, 169, 173, 176, 177, 182, 183, 184, 200, 202, 203, 204, 205, 209
Holding point ... 13, 43, 45, 46, 54, 76, 77, 78, 79, 81, 82, 83, 84, 90, 117, 136, 137, 149, 151, 152, 160, 163, 200, 205, 209, 210, 211, 212
Holding voltage ... 19, 49, 76, 83-84, 120, 149, 151, 152, 153, 169, 173, 174, 183, 205
Hook collector transistor ... 38, 40, 69, 71
Inverter circuit ... 9, 10, 11, 13, 14, 25, 52, 124, 126, 128, 160, 173, 176, 182
Inverter cross section ... 8, 10
Latched state ... 12, 23, 24, 54, 57, 90, 108, 112, 115, 118, 120, 121, 124, 129, 131, 135, 136, 149, 150, 151, 152, 153, 161, 163, 197, 200, 201, 204, 210
Lifetime reduction ... 4, 6, 174-175, 182, 206
Multiple well contacts ... 172, 191, 196, 206
Negative differential resistance ... 11, 43, 45, 53, 54, 118, 124, 125, 129, 134, 136, 138, 139, 150, 151, 154, 163, 203, 205, 209, 210, 211, 212
Parameter analyzer ... 124-125, 129, 130, 131, 133, 138, 140, 153, 202, 209, 213

# Index

PNPN
- characteristics ... 11, 12, 19, 21, 33, 119, 121, 122, 125, 129, 134, 136, 137, 150, 152, 154, 201, 202, 210
- cross section ... 14, 54, 95, 103
- diode ... 17, 41, 42, 44, 45, 48, 51, 58, 119, 120, 126, 202
- loop gain ... 17, 21, 24, 49, 115, 163, 197, 210, 215
- lumped element model ... 7, 14, 15, 16, 37, 38, 47, 84, 85, 90, 93, 98, 99, 103, 151, 198, 200
- tetrode ... 21, 45, 51, 53, 57, 58, 72, 73, 74, 76, 91, 92, 116, 119, 120, 122, 126, 132, 133, 136, 138, 143, 158, 199, 202, 203
- triode ... 17, 21, 34, 42, 43, 44, 45, 51, 69, 71, 72, 73, 74, 76, 92, 93, 116, 119, 120, 122, 123, 126, 133, 139, 143, 144, 146, 147, 163, 194, 199, 200, 202

Retarding base field ... 50, 175-177, 185
Retrograde well ... 4, 26, 50, 175, 176, 177, 181, 184-185, 196, 206, 207
SAFE space ... 7, 23, 66, 67, 69, 71, 84, 90, 91, 116, 117, 157, 165, 194, 199, 200, 201, 202, 207
Saturation region ... 20, 76-84
Schottky source/drain ... 4, 123, 178-180, 206
Semiconductor controlled rectifier (SCR) ... 26, 40, 198
Structural origins of latchup ... 1, 3, 7, 197
Substrate bias ... 25, 28, 31, 53, 87, 156, 187
Substrate contact ring ... 96, 156, 171, 172- 173, 206, 207
Supply overvoltage stress ... 32, 120, 121, 126-129, 130, 133, 134, 139, 152, 176, 186, 192, 193, 201, 202, 204
Switching boundary ... 37, 84, 90, 91, 157, 175, 199, 200, 201, 202
Switching current ... 12, 18, 19, 24, 34, 45, 51, 63, 68, 70, 71, 74, 76, 83, 86, 88, 90, 92, 93, 98, 107, 109, 111, 116, 117, 129, 132, 133, 134, 136, 137, 138-149, 153, 156, 160, 163, 177, 183, 197, 200, 201, 203, 204, 205, 209, 210, 212
Switching point ... 45, 46, 60, 66, 68, 69, 70, 71, 74, 80, 84, 87, 90, 92, 107, 109, 117, 124, 125, 133, 134, 136, 137, 138-149, 163, 200, 204, 210, 211, 214
Temperature dependence of latchup ... 44, 67, 68, 70, 71, 160-161, 204
Topside substrate contact ... 96, 100, 103, 104, 140, 146, 147, 182
Transfer resistance ... 94, 95, 97, 98, 99-104

Transmission line model ... 37, 94-99, 101, 103, 191, 194, 198
Trench isolation ... 4, 5, 102, 188-190, 207
Triggering current ... 188
Triggering modes
    avalanching N-well junction ... 26, 36, 202
    avalanching source/drain junction ... 33, 36
    displacement current ... 33, 34, 36, 118, 127, 155, 203
    input node overshoot/undershoot ... 26, 34, 36, 158, 204, 206
    output node overshoot/undershoot ... 25, 34, 36, 158, 204, 206
    parasitic field device ... 30, 36, 128, 176, 183, 188, 202
    photocurrent ... 32, 34, 36, 162, 181
    external punchthrough ... 28, 36, 128, 130, 176, 202
    internal punchthrough ... 29, 30, 36, 128, 130, 142, 183
Triggering taxonomy ... 34, 36
Turn-off current ... 43, 46, 53, 54, 76, 78, 79, 83, 87, 88, 200, 201
Turn-off point ... 12, 43, 46, 53, 54, 78
Ungated triggering ... 120, 126, 201

# ABOUT THE AUTHOR

Ronald R. Troutman, a senior engineer at IBM, has published widely on FET device design, subthreshold behavior, hot electron effects and latchup in CMOS.

Dr. Troutman received his B.S. from MIT and his Ph.D. from New York University. Most recently, while on technical sabbatical from IBM, Dr. Troutman was a fellow at the Center for Advanced Engineering Studies at MIT.

Printed in the United States
1072000001B